Basic Chemistry for Water and Wastewater Operators

Basic Chemistry for Water and Wastewater Operators

Darshan Singh Sarai

Library of Congress Cataloging-in-Publication Data has been applied for.

Printed in the United States of America
American Water Works Association
6666 West Quincy Avenue
Denver, CO 80235

ISBN 1-58321-148-9

Printed on recycled paper

Table of Contents

Preface

This book is the result of 30 years of teaching and practical experience in the water and wastewater fields. A strong educational background—I have a masters with honors degree in biology from Punjab University, India, and a doctorate degree from the University of Alberta, Canada, in environmental entomology and I am a board-certified entomologist and a registered sanitarian—was very helpful in completing this work. I spent nine years teaching water and wastewater chemistry at Water and Wastewater Technical School, in Neosho, Missouri, without the help of a good textbook. I felt a serious need for a basic chemistry book tailored just for water and wastewater operating staff who generally have little or no knowledge of chemistry. A large part of this work was initiated at that time.

After teaching, I had the opportunity to work at Water District #1 of Johnson County, Kansas, for more than 20 years. This provided a unique chance to apply my teaching knowledge in the treatment of water because I was in charge of the water treatment and quality control laboratory and responsible for training operating staff. During this time, I also taught part-time at a community college and for the Department of Natural Resources of the State of Missouri. I was compiling material while working both in the field and classroom.

After retiring, I was encouraged by my family, students, and friends to publish a book. While preparing the manuscript, I received a very timely letter from Colin Murcray, Manager of Business and Acquisitions for the American Water Works Association, that asked the question, Are you a walking encyclopedia? The letter was sent to water personnel asking them to publish based on their experience in the water industry.

Most concepts covered in this book are in a basic form. For further details, the reader can refer to the references at the end of the book. A set of simple self-test questions are given at the end of each chapter to encourage the reader to further study the subject matter.

I hope this publication will serve its intended purpose of helping operating staff by giving them pertinent chemistry information, which will help them comply with US Environmental Protection Agency regulations. I sincerely acknowledge the help of all those who made this work possible.

1 Introduction

Chemistry is a science that deals with the structure, the composition, and changes in composition of matter as well as the laws that govern these changes. Knowledge of basic chemistry concepts is essential for a person in the water and wastewater field to understand chemical phases of treatment such as coagulation, sedimentation, softening, disinfection, and chemical removal of the various undesirable substances.

Matter and Energy

Matter

Matter is anything that occupies space and has mass. All materials consist of matter, and mass of a body is its quantity of matter. Mass is responsible for the weight of the body—the larger the mass, the heavier the body. Weight is the measurement of earth's attraction to the body. It changes with the distance of the body from the center of the earth—the greater the distance, the less the body weighs, and vice versa. The mass of the body is constant, but weight changes with altitude. An astronaut in space is weightless, although he has the same mass as on earth. For practical purposes, mass and weight are used interchangeably.

Three States of Matter

All matter exists in one or more of three states: solid, liquid, and/or gas. In the solid state, matter has a definite volume and a definite shape (e.g., a block of wood). In the liquid state, there is only definite volume (e.g., water in a glass). In the gas state, there is neither a definite volume nor a definite shape (e.g., air). Some substances can exist in nature in all three states. Water, for example, can be a solid (ice), a liquid (water), and a gas (vapor).

In many instances, the physical state of a substance transforms with a change in temperature: iron into liquid at 3,000°C, ice into water at 0°C, and water into steam at 100°C. At very high temperatures, gases change to a fourth rather rare state of matter called plasma.

Changes in Matter

There are two types of changes in matter—physical and chemical. A physical change is a change in the physical state, such as from solid to liquid and liquid to gas, or ice into water and water into steam. A chemical change is a change in the chemical composition of a substance and, therefore, in its properties, such as the burning of wood to form carbon dioxide, water, and ashes.

Density and Specific Gravity

Density of a substance is the mass per unit volume. A unit of mass in chemistry is a gram and a unit of volume is a cubic centimetre. For gases, it is grams per litre at standard temperature (0°C) and standard pressure (760 mm mercury). The formula for calculating density is

$$D = \frac{M}{V}$$

where:

D = density
M = mass in grams
V = volume as cubic centimetres or litres

Sample Problems

1. 10 cm^3 of iron weighs 78.6 g. What is the density of iron?

$$\text{Density of iron} = \frac{78.6 \text{ g}}{10 \text{ cm}^3}$$
$$= 7.86 \text{ g/cm}^3$$

2. 100 L of oxygen at STP (standard temperature and pressure) measures 143 g. Calculate the density of oxygen.

$$\text{Density of oxygen} = \frac{143 \text{ g}}{100 \text{ L}}$$
$$= 1.43 \text{ g/L}$$

Specific gravity is the ratio of the density of a substance to that of a standard. The standard for solids and liquids is water with the density of 1 g/cm^3. For gases, the standard is air with the density of 1.29 g/L.

For solids and liquids,

$$\text{specific gravity} = \frac{\text{density of the substance}}{\text{density of water}}$$

For gases,

$$\text{specific gravity} = \frac{\text{density of the gas}}{\text{density of air}}$$

3. Calculate the specific gravity of gold with the density 19.3 g/cm^3.

$$\text{Specific gravity of gold} = \frac{19.3 \text{ g/cm}^3}{1 \text{ g/cm}^3}$$
$$= 19.3$$

4. Calculate the specific gravity of oxygen gas with the density 1.43 g/L.

$$\text{Specific gravity of oxygen} = \frac{1.43 \text{ g/L}}{1.29 \text{ g/L}} = 1.11$$

Energy

Energy is the capacity to do work. Common forms of energy include heat, light, and electricity, as well as chemical and mechanical energy. Two types of mechanical energy are kinetic and potential. Energy due to motion is known as kinetic energy, e.g., the energy from wind or water turning the wheel of a mill. Energy due to position is known as potential energy, e.g., falling water in a dam creates energy to turn a turbine.

Law of Conservation of Energy

Energy can be transformed from one form to another; it is neither created nor destroyed. The total amount of energy in the universe is constant.

Einstein has proved through his famous equation, $E = mc^2$, where "E" stands for energy, "m" for mass, and "c" for a constant (speed of light), that energy can be converted into mass and mass into energy. Therefore, the above given law can be written as *the total amount of energy or matter in the universe is constant.*

Properties

Properties of a substance can be classified into three categories:

1. *Physical properties:* The chemical composition of a substance in these properties is unchanged and includes color, taste, odor, solubility, density, hardness, and melting and boiling points.

2. *Chemical properties:* These properties are determined by changing the chemical composition (e.g., through interaction with other substances, burning, and reaction to heat and light).
3. *Specific properties:* These properties are used for the identification of a substance.

Questions

1. Why is chemistry important in water treatment?
2. What is matter?
3. Distinguish between (a) mass and weight, and (b) density and specific gravity.
4. Calculate the density of zinc when 20 cm^3 of it weighs 142.8 g.
5. What is the density of hydrogen if 100 L weighs about 9 g?
6. Define chemical and physical properties.
7. Define a chemical change and give examples of a chemical change and a physical change.
8. Define energy and state the law of conservation of energy.

2 Measurements in Chemistry

The Metric System

The English system of measurements has the disadvantage of lacking a simple numerical relationship between the units and their parts. To overcome this disadvantage, the metric system was developed in France near the end of the eighteenth century. Fractional and multiple parts of the metric basic unit have a simple numerical relationship based on the number 10.

This system is commonly used all over the world in scientific works.

Metric Units

A gram is the basic unit of mass. Originally, a gram was defined as the mass of 1 cm^3 of water at 4°C. Now, it is 1/1,000 of the mass of a standard kilogram weight kept at the International Bureau of Weights and Measures near Paris, France. One gram is equal to 0.0022 lb.

A metre, the basic unit of length, was originally defined as 1/10,000,000 of the distance from the North Pole to the equator along the Paris Meridian. Today, it is the distance between two parallel lines engraved on a platinum-iridium bar kept at the International Bureau of Weights and Measures. One metre is equal to 39.37 in.

A litre is the unit of volume. It is the volume of 1 kg of water at 4°C. Thus, 1 g of water equals 1 mL or 1 cm^3. One litre is equal to 0.264 gal.

Prefixes are used with these units to complete the system. Multiple or ascending values use Greek prefixes:

Deka-, D = 10 basic units
Hecto-, H = 100 basic units
Kilo-, K = 1,000 basic units
Mega-, M = 1,000,000 basic units
Giga-, G = 1,000,000,000 basic units
Tera-, T = 1,000,000,000,000 basic units

Fractional or descending values are identified by Latin prefixes:

Deci-, d = 1/10 of basic unit
Centi-, c = 1/100 of basic unit
Milli-, m = 1/1,000 of basic unit
Micro-, µ = 1/1,000,000 of basic unit
Nano-, n = 1/1,000,000,000 of basic unit
Pico-, p = 1/1,000,000,000,000 of basic unit
Femto-, f = 1/1,000,000,000,000,000 of basic unit
Atto-, a = 1/1,000,000,000,000,000,000 of basic unit

By using a metre as an example, these prefixes are used as follows:

Kilometre = 1,000 metres
= 100 dekametres
= 10 hectometres

Hectometre	= 100 metres
Dekametre	= 10 metres
Metre	
Decimetre	= 1/10 metre
Centimetre	= 1/100 metre
Millimetre	= 1/1,000 metre
	= 1/100 decimetre
	= 1/10 centimetre

Temperature in the metric system is measured on the centigrade (Celsius) scale, which has two fixed points: 0°C, the freezing point of water, and 100°C, the boiling point of water at standard atmospheric pressure. The scale between these two points is divided into 100 equal intervals, the centigrade degrees.

The Fahrenheit scale was devised for practical use in everyday life. The freezing point of water is 32°F, and the boiling point is 212°F; thus, there are 180 units (212 – 32) between these fixed points. Therefore, 1 unit on the centigrade scale is equal to $\frac{9}{5}$ unit on the Fahrenheit scale, and 1 Fahrenheit unit is equal to $\frac{5}{9}$ of a centigrade unit.

	Centigrade (°C)	Fahrenheit (°F)	Absolute (°A)
Boiling point of water	100	212	373
	75	167	348
	50	122	323
	25	77	298
Freezing point of water	0	32	273

Conversion formulae:

$$°F = (°C \times 9/5) + 32$$
$$°C = (°F - 32) \times 5/9$$

The absolute (Kelvin) scale is used for the relationship of the volume of a gas and its temperature. The interval between the freezing and boiling points as on the centigrade scale is 100. The freezing point on the absolute scale is 273°A and the boiling point is 373°A.

$$°A = °C + 273 \text{ or } °C = °A - 273$$

Sample Problems

1. Convert 35°C to degrees Fahrenheit.

   Degrees F $= (35 \times 9/5) + 32$
   $= 95$

1. Convert 68°F to degrees centigrade.

   Degrees C $= (68 - 32) \times 5/9$
   $= 20$

2. Convert 290°A to degrees centigrade.

   Degrees C $= 290 - 273$
   $= 17$

Exponential or Scientific Notation

Scientific notation is a shorthand method for writing very large or very small numbers, such as the distance of the earth from the sun (93,000,000 mi), or the weight of a proton (0.000,000,000,000,000,000,000,001,673 g). Exponential notation is expressed as

$$M \times 10^n$$

where:

M = a number 1 to 9.99

n = exponent or the power of the base 10 (n is positive for numbers higher than 10 and negative for numbers less than 1)

The power of 10 is the number of places the decimal point is moved to have only one digit to the left. If the decimal point is moved to the left, the power is positive and if moved to the right, it is negative. Exponential notation for 93,000,000 mi is 9.3×10^7 mi and for 0.000,000,000,000,000,000,000,001,673 g it is 1.673×10^{-24} g.

Questions

1. Define gram, litre, metre, and 0°C.
2. Give the basic number of the metric system.
3. How many millimetres are in 10 km?
4. Convert the following temperatures to centigrade:
   a. 113°F
   b. 50°F
   c. –40°F
5. Convert the following temperatures to Fahrenheit:
   a. 30°C
   b. –40°C
   c. 50°C
6. Make the following conversions into absolute degrees:
   a. 95°F
   b. 30°C
   c. 104°F
7. Write the following numbers in exponential notation form:
   a. 30,000,000,000 cm/sec, the speed of light
   b. 0.000,000,000,5 cm
   c. 602,200,000,000,000,000,000,000.0 particles, the Avogadro number

3 Elements

Elements are basic substances, such as carbon, oxygen, nitrogen, and chlorine, that singly or in combination constitute all matter. At present, there are 112 known elements; 92 are naturally occurring and the remaining are synthesized.

An *element* is defined as a chemical substance that cannot be decomposed into simpler substances by ordinary chemical changes. The smallest particle of an element that can exist and retain the chemical properties of that element is known as an *atom*.

Atoms are ordinarily unbreakable, but they can be split by using special techniques from the atomic energy field. After splitting, an atom does not have the original properties of its element. It becomes an atom of a lighter element. This process is known as *fission*.

Elements are generally divided into two general classes: metals and nonmetals. Metals are mostly solids, such as gold, silver, copper, and iron. Mercury, a liquid metal, is an exception. They have luster, are good conductors of electricity and heat, and are mostly malleable (can be hammered into sheets) and ductile (can be drawn into wires). Nonmetals are solids or gases. Carbon, sulfur, and phosphorus are solids; whereas, oxygen, fluorine, and chlorine are gases. Bromine is an example of a liquid nonmetal. Solid nonmetals are brittle and dull and are poor conductors of heat and electricity.

Hydrogen is the lightest element, and oxygen is the most abundant element in the earth's crust.

Chemical Symbols

Each element is assigned a symbol to avoid the writing of its name. Symbols are written in English letters. The symbol is either the first letter of the name of the element or the first letter and a second well-pronounced letter from the name. The second letter is used when more than one element starts with the same letter. The first letter of a symbol is capitalized, and the second letter, when present, is lowercased. In several cases, the symbols come from foreign languages, such as Latin, Greek, and German. See Table 3-1 for a partial list of elements and their symbols.

Element	Symbol
Aluminum	Al
Arsenic	As
Bromine	Br
Calcium	Ca
Carbon	C
Chlorine	Cl
Chromium	Cr
Cobalt	Co
Copper	Cu
Fluorine	F
Gold	Au (from Latin name *Aurum*)
Hydrogen	H
Iron	Fe (from Latin name *Ferrum*)
Magnesium	Mg
Manganese	Mn
Nitrogen	N
Oxygen	O

Table 3-1 Symbols for Some Common Elements

Atomic Structure

Atoms are very small particles that cannot be seen even with an ordinary microscope. Atomic diameter ranges from 1–5 angstroms (Å) (an angstrom is 1/100,000,000 of a centimetre, 10^{-8} cm). An atom consists of two main parts: a positively charged central part, the *nucleus*, and a negatively charged outer part, the *shells*, or energy levels. The nucleus is about 1/100,000 of an angstrom and it contains two kinds of particles, positively charged protons and uncharged neutrons. A proton has a mass of 1.673×10^{-24} g, which is approximately the mass of a hydrogen atom. A neutron weighs almost the same as a proton (1.675×10^{-24} g); thus, a proton or a neutron weigh about as much as a hydrogen atom. Shells are the orbits of electrons, which are very small particles revolving around the nucleus. An electron carries a negative charge. The mass of an electron is 9.110×10^{-28} g. It is an insignificant mass as compared to a proton or a neutron. The number of protons in an uncombined atom is equal to the number of electrons. An atom is, therefore, electrically neutral. The *atomic number* of an element is the number of protons in its atom. Some elements may have atoms with different numbers of neutrons; these atoms are known as *isotopes*. Most elements exist as a number of isotopes. Hydrogen, for example, has three isotopes: protium, deuterium, and tritium, with 0, 1, and 2 neutrons, respectively.

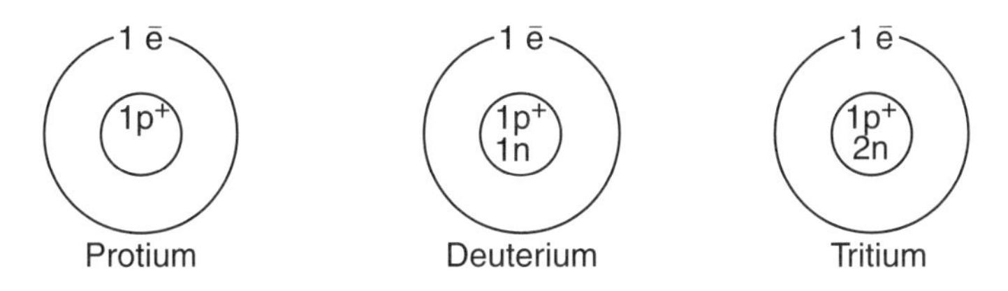

Isotopes of Hydrogen

The *mass number* of an isotope is the sum of its protons and neutrons. It is termed "mass number" because protons and neutrons contribute atomic mass. It is written after the name of the element, e.g., hydrogen-1, hydrogen-2, oxygen-16, and carbon-14. In symbolic form, the oxygen-16 isotope is $_8O^{16}$, where 16 is the mass number and 8 is the atomic number. In some books, it is shown simply as O^{16}.

Atomic Weight

Atoms are too small to be weighed individually; therefore, a system of relative masses was developed in which C^{12} (carbon isotope with mass number 12) is assigned a mass of 12 atomic mass units (AMU) and is used as a standard for atomic weights of all other elements. The *atomic weight* of an element is, therefore, defined as the average mass of its isotopes as compared to a C^{12} atom, which equals 12 AMU. The atomic mass unit is a very small mass.

$$1 \text{ g} = 6.022169 \times 10^{23} \text{ AMU}$$

Avogadro Number and Mole

The Avogadro number is a constant in chemistry. It is 6.022169×10^{23} particles of a substance. (See its use in above equation.) The amount of a substance (an element or a compound) in grams containing an Avogadro number of its particles is known as a *mole*. Whenever grams replace atomic mass units of the mass of a particle, the number of particles changes from 1 to 6.02×10^{23} particles. Thus, the atomic weight of an element with mass units in grams is a mole of that element and is also known as *gram atomic weight*.

$$1 \text{ hydrogen atom} = 1.008 \text{ AMU}$$

$$6.02 \times 10^{23} \text{ hydrogen atoms} = 1.008 \text{ g}$$

Therefore, 1.008 g is a mole or a gram atomic weight of hydrogen. Similarly, 12.0111 g is a mole of carbon, and 15.9994 g is a mole of oxygen.

Periodic Table

The periodic table (see Tables 3-2 and 3-3) is a chart of the elements arranged according to their atomic numbers. Each element is assigned a block. A horizontal row of blocks on the table is called a *period* or *series*. Periods are indicated by Arabic numerals (1,2,3—or 7). A vertical column, called a *group* or *family*, is indicated by a Roman numeral (I, II—or VIII), which represents the electrons in the outermost shell.

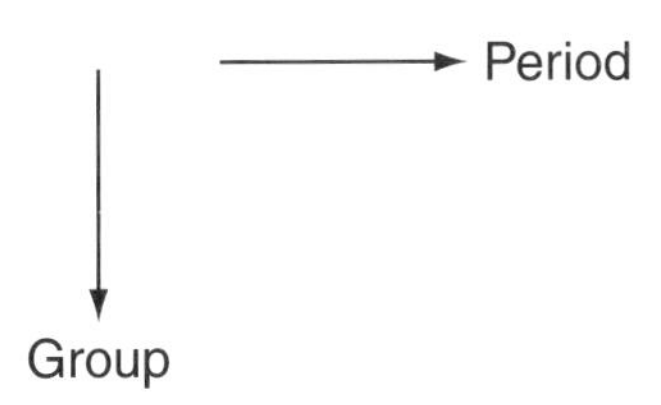

The number of periods indicates the number of shells in its elements. There are seven shells and, thus, seven periods. Elements of the first period (hydrogen and helium) have one shell; elements of the second period (lithium through neon), two shells; and, thus, elements of the seventh period, seven shells. The seven shells are known as K through Q, or one through seven from inside outward. Each shell can hold only a certain number of electrons. K shell can hold a maximum of 2 electrons; L shell, 8; and M shell, 18. Most of the shells are further divided into sublevels identified by the letters s, p, d, and f in order of ascending energy. The s sublevel can hold a maximum of one pair; p, three pairs; d, five pairs; and f, seven pairs of electrons. The path of a pair of electrons is known as an *orbital*. Thus, s sublevel has one orbital; p, three; d, five; and f, seven orbitals. Table 3-4 shows the electronic shells (levels) and sublevels and the maximum number of electrons in each shell.

IA	IIA	IIIB	IVB	VB	VIB	VIIB	VIII			IB	IIB	IIIA	IVA	VA	VIA	VIIA	Inert Gases
1 **H** 1.00797																1 **H** 1.00797	2 **He** 4.0026
3 **Li** 6.939	4 **Be** 9.0122											5 **B** 10.811	6 **C** 12.0112	7 **N** 14.0067	8 **O** 15.9994	9 **F** 18.9984	10 **Ne** 20.183
11 **Na** 22.9898	12 **Mg** 24.312											13 **Al** 26.9815	14 **Si** 28.086	15 **P** 30.9738	16 **S** 32.064	17 **Cl** 35.453	18 **Ar** 39.948
19 **K** 39.102	20 **Ca** 40.08	21 **Sc** 44.956	22 **Ti** 47.90	23 **V** 50.942	24 **Cr** 51.996	25 **Mn** 54.9380	26 **Fe** 55.847	27 **Co** 58.9332	28 **Ni** 58.71	29 **Cu** 63.54	30 **Zn** 65.37	31 **Ga** 69.72	32 **Ge** 72.59	33 **As** 74.9216	34 **Se** 78.96	35 **Br** 79.909	36 **Kr** 83.80
37 **Rb** 85.47	38 **Sr** 87.62	39 **Y** 88.905	40 **Zr** 91.22	41 **Nb** 92.906	42 **Mo** 95.94	43 **Tc** (99)	44 **Ru** 101.07	45 **Rh** 102.905	46 **Pd** 106.4	47 **Ag** 107.870	48 **Cd** 112.40	49 **In** 114.82	50 **Sn** 118.69	51 **Sb** 121.75	52 **Te** 127.60	53 **I** 126.904	54 **Xe** 131.30
55 **Cs** 132.905	56 **Ba** 137.34	* 57 **La** 138.91	72 **Hf** 178.49	73 **Ta** 180.948	74 **W** 183.85	75 **Re** 186.2	76 **Os** 190.2	77 **Ir** 192.2	78 **Pt** 195.09	79 **Au** 196.967	80 **Hg** 200.59	81 **Tl** 204.37	82 **Pb** 207.19	83 **Bi** 208.980	84 **Po** (210)	85 **At** (210)	86 **Rn** (222)
87 **Fr** (223)	88 **Ra** (226)	‡ 89 **Ac** (227)	104 **Rf** (261)	105 **Db** (262)	106 **Sg** (266)	107 **Bh** (262)	108 **Hs** (265)	109 **Mt** (266)	110 **?** (271)	111 **?** (272)	112 **?** (277)						

* Lanthanide Series

58 **Ce** 140.12	59 **Pr** 140.907	60 **Nd** 144.24	61 **Pm** (147)	62 **Sm** 150.35	63 **Eu** 151.96	64 **Gd** 157.25	65 **Tb** 158.924	66 **Dy** 162.50	67 **Ho** 164.930	68 **Er** 167.26	69 **Tm** 168.934	70 **Yb** 173.04	71 **Lu** 174.97

‡ Actinide Series

90 **Th** 232.038	91 **Pa** (231)	92 **U** 238.03	93 **Np** (237)	94 **Pu** (242)	95 **Am** (243)	96 **Cm** (247)	97 **Bk** (247)	98 **Cf** (249)	99 **Es** (254)	100 **Fm** (253)	101 **Md** (256)	102 **No** (256)	103 **Lr** (257)

Numbers in parenthesis are mass numbers of most stable or most common isotope.

Atomic weights corrected to conform to the 1963 values of the Commission on Atomic Weights.

The group designations used here are the former Chemical Abstract Service numbers.

Table 3-2 Periodic Table

Name	Symbol	Atomic Number	Atomic Weight
Actinium	Ac	89	227*
Aluminum	Al	13	26.98
Americium	Am	95	243*
Antimony	Sb	51	121.75
Argon	Ar	18	39.95
Arsenic	As	33	74.92
Astatine	At	85	210*
Barium	Ba	56	137.34
Berkelium	Bk	97	247*
Beryllium	Be	4	9.01
Bismuth	Bi	83	208.98
Bohrium	Bh	107	262
Boron	B	5	10.81
Bromine	Br	35	79.90
Cadmium	Cd	48	112.40
Calcium	Ca	20	40.08
Californium	Cf	98	249*
Carbon	C	6	12.01
Cerium	Ce	58	140.12
Cesium	Cs	55	132.91
Chlorine	Cl	17	35.45
Chromium	Cr	24	52.00
Cobalt	Co	27	58.93
Copper	Cu	29	63.55
Curium	Cm	96	247*
Dubnium	Db	105	262
Dysprosium	Dy	66	162.50
Einsteinium	Es	99	254*
Erbium	Er	68	167.26
Europium	Eu	63	151.96
Fermium	Fm	100	253*
Fluorine	F	9	19.00
Francium	Fr	87	223*
Gadolinium	Gd	64	157.25
Gallium	Ga	31	69.72
Germanium	Ge	32	72.59
Gold	Au	79	196.97
Hafnium	Hf	72	178.49
Hassium	Hs	108	265
Helium	He	2	4.00
Holmium	Ho	67	164.93
Hydrogen	H	1	1.01
Indium	In	49	114.82
Iodine	I	53	126.90
Iridium	Ir	77	192.22
Iron	Fe	26	55.85
Krypton	Kr	36	83.80
Lanthanum	La	57	138.91
Lawrencium	Lr	103	257*
Lead	Pb	82	207.2
Lithium	Li	3	6.94
Lutetium	Lu	71	174.97
Magnesium	Mg	12	24.31
Manganese	Mn	25	54.94
Meitnerium	Mt	109	265
Mendelevium	Md	101	256*

* Mass number of most stable or best-known isotope.

† Mass of most commonly available, long-lived isotope.

Table 3-3 List of Elements

Table continues next page

Name	Symbol	Atomic Number	Atomic Weight
Mercury	Hg	80	200.59
Molybdenum	Mo	42	95.94
Neodymium	Nd	60	144.24
Neon	Ne	10	20.18
Neptunium	Np	93	237.05†
Nickel	Ni	28	58.71
Niobium	Nb	41	92.91
Nitrogen	N	7	14.01
Nobelium	No	102	254*
Osmium	Os	76	190.2
Oxygen	O	8	16.00
Palladium	Pd	46	106.4
Phosphorus	P	15	30.97
Platinum	Pt	78	195.09
Plutonium	Pu	94	242*
Polonium	Po	84	210*
Potassium	K	19	39.10
Praseodymium	Pr	59	140.91
Promethium	Pm	61	147*
Protactinium	Pa	91	231.04*
Radium	Ra	88	226.03†
Radon	Rn	86	222*
Rhenium	Re	75	186.2
Rhodium	Rh	45	102.91
Rubidium	Rb	37	85.47
Ruthenium	Ru	44	101.07
Rutherfordium	Rf	104	261
Samarium	Sm	62	150.4
Scandium	Sc	21	44.969
Seaborgium	Sg	106	263
Selenium	Se	34	78.96
Silicon	Si	14	28.09
Silver	Ag	47	107.87
Sodium	Na	11	22.99
Strontium	Sr	38	87.62
Sulfur	S	16	32.06
Tantalum	Ta	73	180.95
Technetium	Tc	43	98.91†
Tellurium	Te	52	127.60
Terbium	Tb	65	158.93
Thallium	Tl	81	204.37
Thorium	Th	90	232.04†
Thulium	Tm	69	168.93
Tin	Sn	50	118.69
Titanium	Ti	22	47.90
Tungsten	W	74	183.85
Uranium	U	92	238.03
Vanadium	V	23	50.94
Xenon	Xe	54	131.30
Ytterbium	Yb	70	173.04
Yttrium	Y	39	88.91
Zinc	Zn	30	65.38
Zirconium	Zr	40	91.22

* Mass number of most stable or best-known isotope.
† Mass of most commonly available, long-lived isotope.

Table 3-3 List of Elements (continued)

Shell	Sublevels				Maximum Number of Electrons in Shell
	s	p	d	f	
K	2				2
L	2	6			8
M	2	6	10		18
N	2	6	10	14	32
O	2	6	10	14	32
P	2	6	10*		8 or 18
Q	2	6*			2 or 8

* Only sometimes present.

Table 3-4 Shells, Their Sublevels, and Maximum Number of Electrons

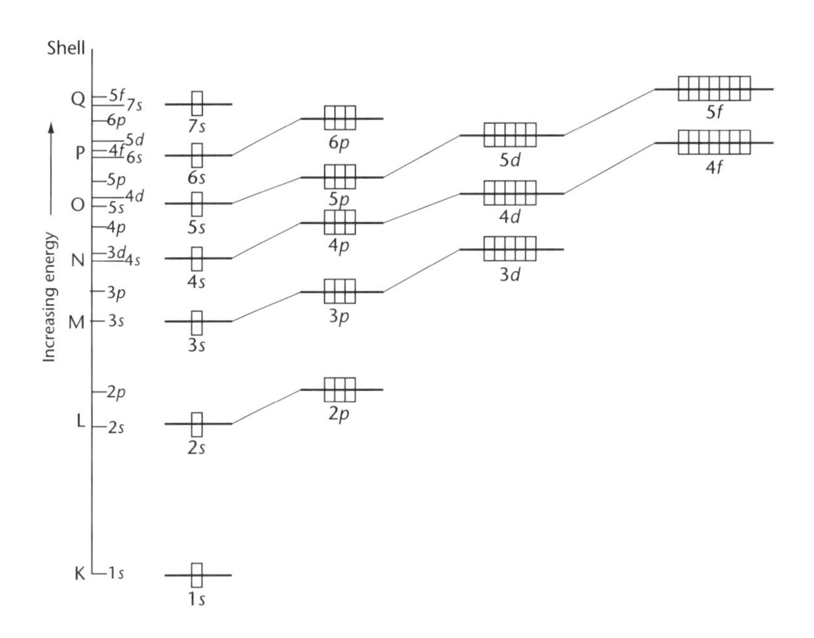

Figure 3-1 Electronic Configuration Based on Energy Level (from *Modern Chemistry* by Metcalf, Williams, and Castka 1966. Publisher: Holt Rinehart & Winston, Inc.)

The electronic arrangement of an atom depends on the energy level of various electrons. They are mostly arranged in such a way that the electrons with minimum energy are closest to the nucleus (in K shell) and with maximum, farthest away. Energy increases from K shell to Q and from s sublevel to f; thus, both the shells and the sublevels are to be considered in determining the place of an electron around the nucleus. Figure 3-1 shows the relative energy levels of electrons in various shells and sublevels.

The electronic structure of an element can be determined by arranging electrons from the base up the electronic energy scale. For example, hydrogen, atomic number 1, has only one electron, which will be in the K shell and s sublevel. It is represented as $1s^1$ where coefficient 1 indicates the shell number and superscript 1 indicates the number of electrons in that sublevel. Helium's electronic structure is $1s^2$, lithium is $ls^2 2s^1$, and sodium is $1s^2 2s^2 2p^6 3s^l$. The sum of the superscripts is, therefore, the atomic number of the element. Each period starts with an element with one electron in its outermost shell and ends with the element at a stable state. There are two stable states in the atoms: two electrons in the K shell, when it is the only shell as in the case of the elements in period 1, and eight electrons in the outermost shell, when more than one shell is present. The outermost shell when it is not stable is known as a *valence shell* and its electrons known as *valence electrons*.

Periodic Law

The *chemical properties of elements repeat periodically as the atomic numbers increase*. Chemical properties are determined by the electrons in the outermost shell. Elements in the same group have similar properties. As the atomic number increases, after every certain number of elements, there are the same number of electrons in the outermost shell. For example, after atomic number 3, which is lithium, sodium with atomic number 11 and potassium with atomic number 19 have one electron in their outermost shells, and they all belong to the same group IA and have similar chemical properties.

The group number indicates the number of electrons in the outermost shell. Group IA has one electron, IIA two electrons, IIIA three electrons . . . and VIIIA eight electrons in their outermost shells. Helium, in group VIIIA, is the only element with two electrons, as it has only a K shell, which is in the stable state with two electrons. In the middle of the table, there are 30 elements, known as *transition elements*. They have electrons in the d sublevels of the third, fourth, and fifth shells. The lower two rows of 14 elements each, known as rare earth elements, have electrons in the f sublevels of the fourth and fifth shells. Transition and rare earth elements have one or two electrons in their valence shells.

Atoms of different elements combine to achieve the stable state (two or eight electrons in the outermost shell). Commonly, elements with one to three electrons in their valence shell (e.g., metals) lose electrons to those elements with valence electrons from five to seven as nonmetals to form compounds. Thus, elements in groups IA to IIIA, transition and rare earth elements, lose electrons and are known as metals, and those in groups VA to VIIA gain electrons and are known as nonmetals. A heavy, staircase-like line on the periodic table separates metals from nonmetals. Elements of group VIIIA are in the stable state and ordinarily do not react with other elements and are, therefore, known as inert or noble gases.

Questions

1. Define element, atom, proton, neutron, and electron.
2. What is the difference between a group and a period?

3. Write symbols for:
   a. Iron
   b. Manganese
   c. Carbon
   d. Mercury
   e. Copper
   f. Silver
4. What is the difference between atomic number and atomic weight?
5. Fill in the blanks:

Element	Protons	Atomic Number	Mass Number	Electrons	Neutrons
Carbon	____	6	____	____	6
Oxygen	____	8	17	____	____
Chlorine	17	____	35	____	____

6. a. How many carbon atoms are in exactly 12 g of carbon-12?
   b. What is the name for this number?
   c. What is the term for the amount of a substance containing this number of particles?
7. a. How many shells are in elements in period 6?
   b. How many electrons are in the outermost shell of elements in group VA?
8. What is the name for atoms with the same atomic number, but different mass number?
9. Define valence shell and valence electrons.

4 Compounds

Every substance in the world is either an element, a compound, or a mixture of various compounds and elements. A compound is formed of two or more elements, which are always present in the same definite proportion. For example, water is formed of 88.89 percent oxygen and 11.11 percent hydrogen by weight.

Law of definite composition: The composition of a chemical compound always contains the same proportions, or a chemical compound has a definite composition by weight. Why do different elements combine to form compounds? They do so to have their outermost shell in the stable state. Electrons are either transferred from the valence shell of one atom to that of another or shared between the valence shells of two atoms to accomplish atomic stability.

Some substances can be mixed together without combining them, such as sugar and salt. The following table shows differences between a compound and a mixture:

Compound	Mixture
1. Has a definite composition by weight.	1. Components may be present in any proportion.
2. Usually, there is evidence of chemical change in a compound formation in the form of heat or light.	2. No such evidence.

Compound	Mixture
3. Components lose their identities and cannot be separated mechanically.	3. Components retain their identities and can be separated by mechanical means.
Examples of compounds are water, carbon dioxide, etc.	Example of a mixture is air, formed mainly of nitrogen and oxygen.

Oxidation number (valence): This number, assigned to each element, indicates the number of electrons lost, gained, or shared by its atom in the compound formation. This number is either positive (+) or negative (–). An atom that loses electrons gains a positive charge and its oxidation number is positive because in an uncombined atom, the number of protons is equal to the number of electrons. The number of electrons lost leaves the atom with the same number of positive charges. Similarly, an atom that gains electrons, gains negative charges equal to the number of electrons gained.

As in the formation of different compounds, the same element may have a different number of electrons involved, so it may have more than one oxidation number. Metals lose electrons and thus have positive oxidation numbers, and nonmetals gain electrons and have negative oxidation numbers. Metals in group IA–IIIA have +1, +2, and +3 oxidation numbers because they lose one, two, and three electrons, respectively. Nonmetals in group VA–VIIA have –3, –2, and –l oxidation numbers because they gain three, two, and one electrons, respectively. The rules for determining oxidation numbers are listed in Table 4-1.

Ionic Compounds

Ionic compounds are formed when a transfer of electrons occurs, which produces ionic or electrovalence

1. For simple ions, the oxidation number is the charge on the ion; i.e.,
$$Na^+ = +1, Cl^- = -1, A1^{+3} = +3$$
2. The oxidation number for a free element is zero; i.e.,
$$H_2 = 0, O_2 = 0, N_2 = 0, Fe = 0, O_3 = 0$$
3. The normal oxidation number of hydrogen is +1 and of oxygen is –2.
4. The algebraic sum of the oxidation numbers within a radical is equal to its charge.
5. The total charge on a chemical compound is zero.

Table 4-1 Rules for Determining Oxidation Numbers

bonding. Ionic bonds are formed when metals combine with nonmetals. The formation of sodium chloride by transferring one electron from a sodium atom to a chlorine atom is an example of ionic bonding (Figure 4-1).

Sodium and chlorine atoms are no longer neutral—the sodium atom has gained a positive charge and the chlorine atom a negative charge. A charged particle is known as an *ion*. A positively charged ion is termed *cation* and a negatively charged ion is *anion*. In ionic compounds, positive and negative ions are held together by electrical forces due to their opposite charge. These forces form an ionic bond. Because only valence shells are involved in the compound formation, they are shown with electrons as dots in a compound formation diagram. For example, sodium chloride and calcium chloride are shown as

Na ∘ ∘Cl (with dots)

Sodium Chloride = NaCl

Cl ∘ Ca ∘ Cl (with dots)

Calcium Chloride = $CaCl_2$

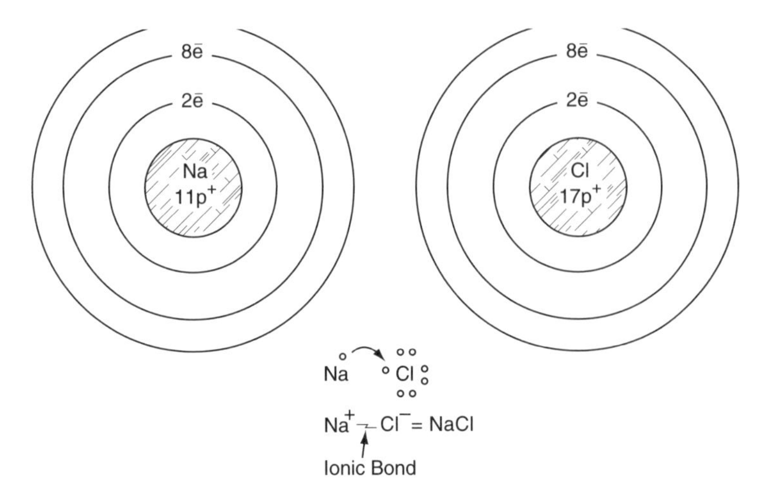

Figure 4-1 Ionic Bonding Between Sodium and Chlorine Atoms

Oxidation is any chemical change in which electrons are lost. In the above examples, it is the oxidation of sodium and calcium atoms. A substance that is oxidized is known as a *reducing agent.*

Reduction is the reverse of oxidation. Reduction is a chemical change in which electrons are gained. In the case of sodium chloride, the chlorine atom has undergone the reduction. A substance that is reduced is known as an *oxidizing agent,* such as chlorine in sodium chloride. Metals are reducing agents and nonmetals are oxidizing agents.

Covalent Compounds

Covalent compounds are formed by sharing of electrons. Sharing of electrons occurs when a nonmetal combines with a nonmetal, or when atoms of the same element combine to form molecules. A *molecule* is the smallest and most stable and neutral particle of a substance. The following are some examples of covalent bonding:

Hydrogen molecule = H_2 = (H ⁝ H) = H–H
covalent bond

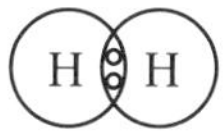

Two hydrogen atoms share a pair of their electrons to form a covalent bond of its molecule.

Water = H_2O = (H ⁝ O ⁝ H) = H–O–H
covalent bonds

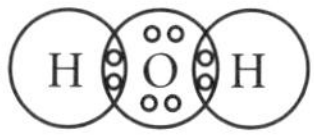

In water formation, two hydrogen atoms share two electrons with an oxygen atom, as shown in the above diagram, and all three atoms are at the stable state. The measure of attraction of an atom for the shared electrons between it and another atom is called *electronegativity*. Maximum electronegativity, which is 4, is assigned to the element fluorine. More electronegativity means more attraction, and vice versa. If the electronegativity difference between two combining atoms is more than 1.6, there is ionic bonding; if less than 1.6, then covalent bonding. When this difference is 0.6–1.6, there is unequal sharing of electrons, which means the atom with more electronegativity shares more than the other. The atom that shares more is assigned a negative oxidation number, and the other a positive number. These compounds are known as *polar covalent compounds* because they carry positive and negative poles on their molecules. (See Table 4-2 for examples of the electronegativity difference and bonding.)

Radicals are charged groups of covalently bonded atoms. They are ions and part of a compound. Most radicals have oxygen bonded with a nonmetal, such as sulfate (SO_4^{-2}), nitrate (NO_3^{-}), carbonate (CO_3^{-2}), and phosphate (PO_4^{-3}). There are some that contain metals, such as permanganate (MnO_4^{-}) and dichromate ($Cr_2O_7^{-2}$). Table 4-3 lists some common ions and their charges (oxidation

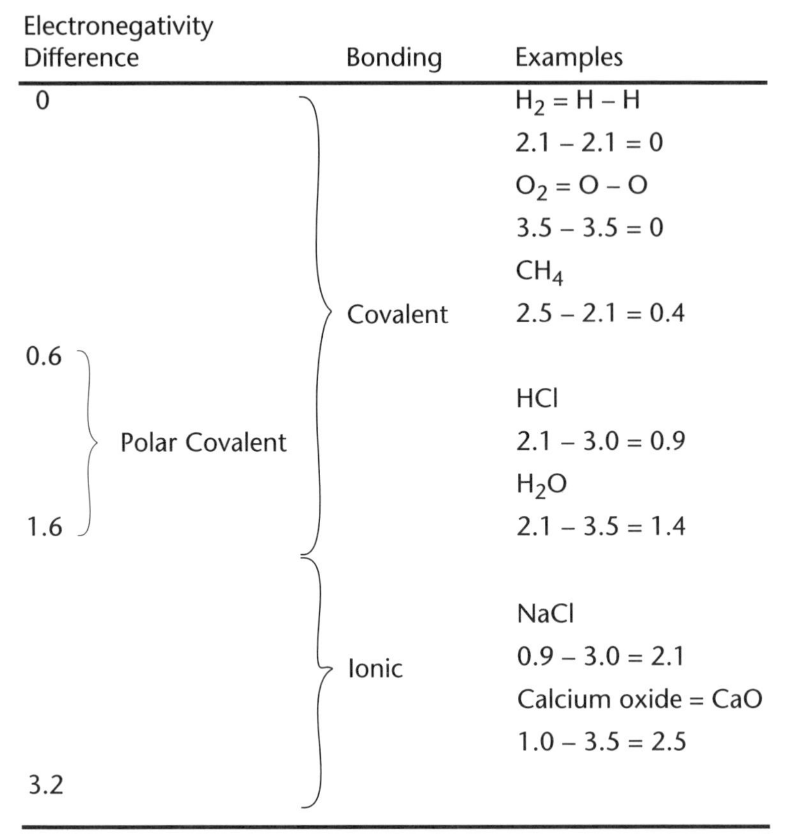

Electronegativity Difference		Bonding	Examples
0		Covalent	$H_2 = H - H$
			2.1 – 2.1 = 0
			$O_2 = O - O$
			3.5 – 3.5 = 0
			CH_4
			2.5 – 2.1 = 0.4
0.6	Polar Covalent		HCl
			2.1 – 3.0 = 0.9
			H_2O
1.6			2.1 – 3.5 = 1.4
		Ionic	NaCl
			0.9 – 3.0 = 2.1
			Calcium oxide = CaO
			1.0 – 3.5 = 2.5
3.2			

Table 4-2 Electronegativity Difference (Absolute) and Chemical Bonding

numbers). The oxidation number of a radical is the algebraic sum of the oxidation number of the atoms in its formula.

Chemical Formulae

A chemical formula is a shorthand notation using chemical symbols and numerical subscripts to represent the chemical composition of a compound. The subscript indicates the number of combined atoms of its respective element. There are three kinds of chemical formulae:

Cations

Oxidation Number					
+1		+2		+3	
Ammonium	NH_4^+	Barium	Ba^{++}	Aluminum	Al^{+++}
Lithium	Li^+	Calcium	Ca^{++}	Iron (III), ferric	Fe^{+++}
Sodium	Na^+	Copper (II), cupric	Cu^{++}		
Potassium	K^+	Iron (II), ferrous	Fe^{++}		
Silver	Ag^+	Lead	Pb^{++}		
Copper (I), cuprous	Cu^+	Magnesium	Mg^{++}		
		Nickel (II)	Ni^{++}		
		Zinc	Zn^{++}		

Anions

Oxidation Number					
–1		–2		–3	
Acetate	$C_2H_3O_2^-$	Carbonate	$CO_3^=$	Phosphate	$PO_4^{\equiv}$
Azide	N_3^-	Chromate	$CrO_4^=$		
Chloride	Cl^-	Dichromate	$Cr_2O_7^=$		
Bromide	Br^-	Oxide	$O^=$		
Iodide	I^-	Peroxide	$O_2^=$		
Hydroxide	OH^-	Sulfate	$SO_4^=$		
Hydrogen Carbonate (Bicarbonate)	HCO_3^-	Sulfide	$S^=$		
Hydrogen Sulfate (Bisulfate)	HSO_4^-	Sulfite	$SO_3^=$		
Nitrate	NO_3^-	Thiosulfate	$S_2O_3^=$		
Nitrite	NO_2^-				
Chlorate	ClO_3^-				
Hypochlorite	ClO^- or OCl^-				

Table 4-3 Some Common Ions and Their Oxidation Numbers

Compound	Empirical Formula	Molecular Formula	Structural Formula
Water	H_2O	H_2O	H–O–H
Ammonia	NH_3	NH_3	H–N–H (with H bonded below N)
Hydrogen peroxide	HO	H_2O_2	H–O–O–H
Methane	CH_4	CH_4	H–C–H (with H bonded above and below C)

Table 4-4 Examples of Formulae

empirical, molecular, and structural. An empirical formula shows relative numbers of atoms, a molecular formula shows the actual number of combined atoms in a molecule, and a structural formula shows the bonding of the atoms. Examples of these formulae are shown in Table 4-4.

In the case of most inorganic compounds (compounds of elements other than carbon), the empirical formula is the same as the molecular formula.

A chemical formula is neutral and formed of two parts: the first part has a positive oxidation number (less electronegativity) and the second part has a negative oxidation number (more electronegativity). When writing the formula of a compound, write the positive part first and the negative part second with their respective oxidation numbers and balance the plus charge with the minus by assigning appropriate subscripts, if required. If both parts of the formula have the same oxidation number (e.g., in sodium chloride, sodium is +1 and chlorine –1), the charge is already balanced, and, therefore, no subscripts are required. When the oxidation numbers of two parts are different, then the charge is commonly balanced by exchanging the oxidation number of the first part for the subscript of the second part, and vice versa. For example, the formula for sodium sulfate, $Na^{+1} SO_4^{-2}$, is Na_2SO_4; two sodium atoms have a +2 charge, and one sulfate radical has a –2 charge. When a radical has a subscript higher than 1, use parentheses and write the subscript outside the parentheses. Study the following examples:

Compound	Oxidation Numbers	Formula
Water	$H^{+1}O^{-2}$	H_2O
Calcium oxide	$Ca^{+2}O^{-2}$	CaO
Calcium hydroxide	$Ca^{+2}OH^-$	$Ca(OH)_2$
Aluminum sulfate	$Al^{+3}SO_4^{-2}$	$Al_2(SO_4)_3$
Ammonium sulfate	$NH_4^+SO_4^{-2}$	$(NH_4)_2SO_4$

Naming Compounds

In most instances, the name of a compound is formed from the name of the first part of the formula followed by the name of the second part. $CaSO_4$ is calcium sulfate. The name of a binary compound (a compound formed of only two elements), however, is formed from the full name of the first element and the root of the name of the second element ending with *-ide*.

Binary Compound	Name
NaCl	Sodium chloride
NaI	Sodium iodide
FeS	Iron sulfide
H_2S	Hydrogen sulfide
CaO	Calcium oxide

When the same two elements form more than one compound, use appropriate prefixes. Carbon and oxygen form CO and CO_2, which are known as carbon monoxide and carbon dioxide, respectively. PBr_5 is known as phosphorus pentabromide, and PBr_3 is phosphorus tribromide.

Formula weight is the sum of weights of all the atoms in a formula. The following table has formula weights of some compounds:

Compound	Formula Weight
Water, H_2O	$(1 \times 2) + 16 = 18$
Methane, CH_4	$12 + (1 \times 4) = 16$
Calcium hydroxide, $Ca(OH)_2$	$40 + (16 \times 2) + (1 \times 2) = 74$

The gram formula weight of a compound is its mole because it contains an Avogadro number of its formulae.

$$1\ H_2O \text{ molecule} = 18 \text{ AMU}$$

$$6.02 \times 10^{23}\ H_2O \text{ molecules} = 18 \text{ g} = 1 \text{ mol of water}$$

The *percentage composition of a compound* is the percentage by mass of various constituent elements in a compound. Percentage is parts per 100 parts. It is a quantitative or stoichiometric relationship of various elements in a compound.

$$\text{Percentage of an element in a compound} = \frac{\text{total atomic weight of the element}}{\text{formula weight}} \times 100$$

Sample Problems

1. Calculate percentage composition of water.

   Formula weight of $H_2O = 18$

   % of hydrogen = $(2/18) \times 100 = 11.11$

   % of oxygen = $(16/18) \times 100 = 88.89$

2. Calculate percentage composition of sodium carbonate, Na_2CO_3.

   Formula weight of $Na_2CO_3 = 106$

   % of Na = $(2 \times 23)/106 \times 100 = 43.4$

   % of C = $(1 \times 12)/106 \times 100 = 11.3$

   % of O = $(3 \times 16)/106 \times 100 = 45.3$

3. Calculate percentage composition of aluminum sulfate, $Al_2(SO_4)_3$.

   Formula weight of $Al_2(SO_4)_3 = 342$

   % of Al = $(2 \times 27)/342 \times 100 = 15.8$

   % of S = $(3 \times 32)/342 \times 100 = 28.1$

   % of O = $(12 \times 16)/342 \times 100 = 56.1$

Questions

1. Define compound, ion, radical, and molecule.
2. Name the two types of bonds in compounds. How are they formed?
3. Write the formulae for the following compounds:
   a. Carbon dioxide
   b. Ammonia
   c. Calcium sulfate
   d. Sodium carbonate
4. Name these compounds:
   a. Na_2O
   b. H_2S
   c. $Ca(OH)_2$
   d. Fe_2O_3
5. Calculate percentage composition of calcium carbonate ($CaCO_3$).

5 Chemical Equations

A chemical equation is a concise symbolized picture of a chemical reaction. It has both qualitative and quantitative significance. It tells us which substances are reacting and in what amounts to produce which and how much of new substances.

$$\text{Hydrogen} + \text{Oxygen} \rightarrow \text{Water}$$

$$2H_2 + O_2 \rightarrow 2H_2O$$

$$\text{Mass Ratio } 2 \times 2 : 2 \times 16 : 2 \times 18$$

This equation shows that 4 mass units of hydrogen react with 32 mass units of oxygen to produce 36 mass units of water. Reacting substances are known as *reactants,* which are written on the left side of the equation. Substances produced are termed *products,* which are written on the right side. An arrow separates reactants from products and indicates the direction of the reaction. Arrows facing both ways indicate a reaction reversing under ordinary experimental conditions.

$$\text{Reactants} \rightleftarrows \text{Products}$$

An arrow facing down ($CaCO_3\downarrow$) indicates precipitation of a product, and an arrow facing up ($CO_2\uparrow$) means a product is a gas.

Law of Conservation of Masses

The law of conservation of masses states that in ordinary chemical reactions, the total mass of reactants is equal to the total mass of products. Only atoms of reactants are rearranged to form products in a chemical reaction. Therefore, if there is the same number of atoms of each element on each side of the equation, the law is satisfied and the equation is balanced.

Steps in Writing a Balanced Equation

1. Write the equation in words for the reaction.
2. Give the correct formulae or symbols for all reactants and products.
3. Balance the equation by assigning appropriate coefficients to reactants and products.
4. Make sure that coefficients are in the smallest whole number ratio.

Here is the reaction of water formation from hydrogen and oxygen, written step-by-step to balance the equation.

1. Hydrogen + Oxygen → Water
2. $H_2 + O_2 \rightarrow H_2O$

   To get two oxygen atoms on the right side, we need two water molecules, which makes four hydrogen atoms on the right side. So we need two hydrogen molecules on the left to balance the equation.
3. $2H_2 + O_2 \rightarrow 2H_2O$
4. 2 : 1 : 2

   Coefficients are in the smallest whole number ratio.

   The elements hydrogen, nitrogen, oxygen, fluorine, chlorine, bromine, and iodine occur in a diatomic (two-atom) molecular state, and all others occur as single atoms. They are written as H_2, N_2, O_2, F_2, Cl_2, Br_2, and I_2 in chemical reactions.

Once these four steps are understood, they can be written in two steps to balance the equation. These two steps are word equation and balanced formula equation.

Examples:

1. Mercury (II)/mercuric oxide → Mercury + Oxygen
   $2HgO \rightarrow 2Hg + O_2\uparrow$
2. Calcium hydroxide + Calcium bicarbonate →
   Calcium carbonate + Water
   $Ca(OH)_2 + Ca(HCO_3)_2 \rightarrow 2CaCO_3\downarrow + 2H_2O$
3. Aluminum sulfate + Calcium hydroxide →
   Aluminum hydroxide + Calcium sulfate
   $Al_2(SO_4)_3 + 3Ca(OH)_2 \rightarrow 2Al(OH)_3\downarrow + 3CaSO_4$

If the same radical is present on both sides of the equation, it is easier to balance it as a single unit. Because $3SO_4^{-2}$ is on the left, make $3SO_4^{-2}$ on the right by using coefficient 3 for $CaSO_4$.

Quantitative Calculations or Stoichiometry in Chemical Reactions

Equations also have quantitative significance. The quantitative relationship of reactants and their products is known as *stoichiometry*. We can calculate amounts of reactants required to produce any desired amount of a product by using the mass ratio of these substances from their formula weights. Let us return to the equation in the previous example where calcium hydroxide reacts with calcium bicarbonate to produce calcium carbonate and water. The balanced equation shows that 1 formula weight of $Ca(OH)_2$ reacts with 1 formula weight of $Ca(HCO_3)_2$ to produce 2 formula weights of $CaCO_3$ and 2 formula weights of H_2O.

Mass ratio 74 : 162 : 200 : 36

$$Ca(OH)_2 + Ca(HCO_3)_2 \rightarrow 2CaCO_3\downarrow + 2H_2O$$

This also means that 1 mol (gram formula weight) of $Ca(OH)_2$ and 1 mol of $Ca(HCO_3)_2$ are needed to produce

2 mol of $CaCO_3$ and 2 mol of H_2O. By converting moles into grams, we learn that 74 g of $Ca(OH)_2$ react with 162 g of $Ca(HCO_3)_2$ to produce 200 g of $CaCO_3$ and 36 g of H_2O.

Using this information, we can calculate the amount of any substance corresponding to the given mass of any other substance in the reaction. Let us calculate the amount of $Ca(OH)_2$ required to produce 100 g of $CaCO_3$. The equation shows that 200 g of $CaCO_3$ will be produced from 74 g of $Ca(OH)_2$. Thus, 100 g of $CaCO_3$ would require 37 g (one half of 74) of $Ca(OH)_2$. Therefore, the mass ratio of $Ca(OH)_2$ to $CaCO_3$ stays the same in both cases.

This calculation can be done systematically in three steps:

1. Balance the equation.
2. Establish the mass ratio of the two substances.
3. From the mass ratio, calculate the mass of one substance corresponding to the given mass of the other as follows:

(3) X   100 g

(2) 74 g   200 g

(1) $Ca(OH)_2 + Ca(HCO_3)_2 \rightarrow 2CaCO_3\downarrow + 2H_2O$

$$\frac{X}{74\text{ g}} = \frac{100\text{ g}}{200\text{ g}}$$

$$X = \frac{100}{200} \times 74\text{ g}$$

$$= 37\text{ g}$$

How many milligrams per litre of calcium oxide are required to react with 5 mg/L of carbon dioxide in water?

(3) X   5 mg

(2) 56 mg 44 mg

(1) $CaO + CO_2 \rightarrow CaCO_3\downarrow$

$$\frac{X}{56\ \text{mg}} = \frac{5\ \text{mg}}{44\ \text{mg}}$$

$$X = \frac{5}{44} \times 56\ \text{mg}$$

$$= 6.36\ \text{mg/L}$$

Various Types of Chemical Reactions

Chemical reactions can be classified into composition, decomposition, replacement, ionic, and redox reactions.

Composition reactions: In these reactions, two or more substances combine to form a new compound. For example, iron and sulfur combine to form iron sulfide:

$$Fe + S \rightarrow FeS$$

Hydrogen and oxygen react to form water:

$$2H_2 + O_2 \rightarrow 2H_2O$$

Sodium and chlorine combine to form sodium chloride:

$$2Na + Cl_2 \rightarrow 2NaCl$$

Decomposition reactions: They are the reverse of the composition reactions. A complex substance breaks down into simpler substances. For example, electrolysis (decomposition by means of electricity) of water produces hydrogen and oxygen:

$$2H_2O \rightarrow 2H_2\uparrow + O_2\uparrow$$

Quicklime (CaO) is formed by heating limestone ($CaCO_3$):

$$CaCO_3 \xrightarrow{\triangle} CaO + CO_2\uparrow$$

$\triangle$ stands for heat

Replacement reactions: In these reactions, an element is replaced in a compound by another element. The stronger element replaces the weaker element. For example, iron replaces copper from a solution of copper sulfate ($CuSO_4$), forming iron sulfate and copper. It happens because iron is a stronger reducing agent than copper.

$$Fe + CuSO_4 \rightarrow FeSO_4 + Cu\downarrow$$

Table 5-1 shows the increasing strength of oxidizing and reducing agents. This table helps to determine which element can replace other elements.

Ionic reactions or double replacement: In these reactions, a part of an ionic compound is exchanged with a part of another when their solutions are mixed together. One set of ions forms an insoluble compound that precipitates out. Following are some examples of ionic reactions from water and wastewater treatment.

In the removal of turbidity, alum ($Al_2(SO_4)_3$) and lime ($Ca(OH)_2$) are mixed with water to form $Al(OH)_3$, the floc:

$$Al_2(SO_4)_3 + 3Ca(OH)_2 \rightarrow 2Al(OH)_3\downarrow + 3CaSO_4$$

In the removal of permanent hardness caused by sulfate, chloride, or nitrate of calcium, soda ash (Na_2CO_3) is used and calcium carbonate ($CaCO_3$) is formed and precipitates out:

$$CaSO_4 + Na_2CO_3 \rightarrow CaCO_3\downarrow + Na_2SO_4$$

Magnesium hardness is removed by using lime ($Ca(OH)_2$):

$$MgSO_4 + Ca(OH)_2 \rightarrow Mg(OH)_2\downarrow + CaSO_4$$

Redox reactions: The term *redox* comes from reduction and oxidation. In these reactions, two elements have different oxidation numbers on two sides of the equation. Any equation with a free element in it is a redox reaction.

Reduced State	Oxidized State	Oxidation Potential, *volts*
Cs	$Cs^{+} + e^{-}$	3.02
Li	$Li^{+} + e^{-}$	3.02
K	$K^{+} + e^{-}$	2.99
Ba	$Ba^{++} + 2e^{-}$	2.90
Ca	$Ca^{++} + 2e^{-}$	2.87
Na	$Na^{+} + e^{-}$	2.71
Mg	$Mg^{++} + 2e^{-}$	2.34
Al	$Al^{+3} + 3e^{-}$	1.67
Mn	$Mn^{++} + 2e^{-}$	1.05
Zn	$Zn^{++} + 2e^{-}$	0.76
Cr	$Cr^{+3} + 3e^{-}$	0.71
Fe	$Fe^{++} + 2e^{-}$	0.44
Co	$Co^{++} + 2e^{-}$	0.28
Ni	$Ni^{++} + 2e^{-}$	0.25
Sn	$Sn^{++} + 2e^{-}$	0.14
Pb	$Pb^{++} + 2e^{-}$	0.13
H_2	$2H^{+} + 2e^{-}$	0.00
Sn^{++}	$Sn^{+4} + 2e^{-}$	–0.15
Cu	$Cu^{++} + 2e^{-}$	–0.34
$2I^{-}$	$I_2 + 2e^{-}$	–0.53
Fe^{++}	$Fe^{3} + e^{-}$	–0.75
Hg	$Hg^{+} + e^{-}$	–0.80
Ag	$Ag^{+} + e^{-}$	–0.80
Hg	$Hg^{++} + 2e^{-}$	–0.85
Hg^{+}	$Hg^{++} + e^{-}$	–0.91
$2Br^{-}$	$Br_2 + 2e^{-}$	–1.06
$2Cl^{-}$	$Cl_2 + 2e^{-}$	–1.36
$2F^{-}$	$F_2 + 2e^{-}$	–3.03

Increasing Strength of Reducing Agent (↑, upward along Reduced State column)

Increasing Strength of Oxidizing Agent (↓, downward along Oxidized State column)

Table 5-1 Electromotive Series

All replacement reactions, therefore, are redox reactions. For example, formation of a salt and hydrogen gas when an acid reacts with a metal is shown as

$$Zn + 2HCl \rightarrow ZnCl_2 + H_2\uparrow$$

The rusting of iron is shown as

$$4Fe + 3O_2 \rightarrow 2Fe_2O_3\downarrow$$

The formation of iodine from sodium iodide (NaI) by chlorine is shown as

$$2NaI + Cl_2 \rightarrow 2NaCl + I_2$$

How to Balance a Redox Equation

Suppose the reaction is

Hydrogen sulfide + Oxygen → Sulfur Dioxide + Water

Step 1: Write the skeleton equation for the reaction.

$$H_2S + O_2 \rightarrow SO_2 + H_2O$$

Step 2: Assign oxidation numbers to all elements and determine which one is oxidized and which one is reduced.

$$\overset{+1-2}{H_2S} + \overset{0}{O_2} \rightarrow \overset{+4-2}{SO_2} + \overset{+1-2}{H_2O}$$

Sulfur, in this case, is oxidized from –2 to +4 and oxygen is reduced from 0 to –2.

Step 3: Write the electronic equations for the oxidation process and reduction process.

$$S^{-2} - 6e^- \rightarrow S^{+4}$$

$$O^0{}_2 + 4e^- \rightarrow 2O^{-2}$$

Step 4: Adjust the coefficients in both electronic equations so that the number of electrons lost equals electrons gained. The smallest number common to both equations is 12.

$$2S^{-2} - 12e^- \rightarrow 2S^{+4}$$

$$3O^0_2 + 12e^- \rightarrow 6O^{-2}$$

Step 5: Use the above coefficients in the skeleton equation and balance the rest of the equation by trial-and-error method.

$$2H_2S + 3O_2 \rightarrow 2SO_2 + H_2O$$

Balanced equation:

$$2H_2S + 3O_2 \rightarrow 2SO_2 + 2H_2O$$

Questions

1. What is a chemical equation?
2. Write an equation for formation of water by burning hydrogen in oxygen.
3. What is the practical use of chemical equations?
4. State the law of conservation of masses.
5. Balance the following equations:

   a. Oxygen $\rightleftarrows$ Ozone
      $O_2 \rightleftarrows O_3$

   b. $Ca(OH)_2 + CO_2 \rightarrow CaCO_3\downarrow + H_2O$

   c. $Ca(OH)_2 + Mg(HCO_3)_2 \rightarrow CaCO_3\downarrow + MgCO_3 + H_2O$

   d. $CaCO_3 + H_2SO_4 \rightarrow CaSO_4 + H_2CO_3$
      Sulfuric acid

   e. $Ca(OCl)_2 + H_2O \rightarrow Ca(OH)_2 + HOCl$
      Calcium hypochlorite   Hypochlorous acid

   f. $NH_3 + HOCl \rightarrow NHCl_2 + H_2O$

6. Balance the following equation and calculate the amount of oxygen required to react with 5 g of hydrogen sulfide.

$$H_2S + O_2 \rightarrow H_2SO_4$$

6 Acids, Bases, and Salts

Compounds whose water solutions contain ions are acids, bases, or salts.

Acids

An *acid* is a substance that gives hydrogen ions (protons) in its water solution. According to the more general modern concept, an acid is a substance that donates protons to another substance.

A hydrogen ion (H^+) is known as a proton because there is a single proton in the hydrogen nucleus that gives a positive charge after the loss of the electron. A hydrogen ion in a water solution exists only in association with a water molecule as hydronium ion (H_3O^+).

$$H^+ + H_2O = H_3O^+$$

$$HCl + H_2O \rightarrow H_3O^+ + Cl^-$$

Sometimes, this equation is simplified as

$$HCl \rightarrow H^+ + Cl^-$$

A hydronium ion is responsible for the properties of the acids, such as sour taste and the ability to turn blue litmus paper red.

All acids are polar covalent compounds; hydrogen, the common element in all acids, is commonly the first part of their formula. In general, they are water solutions.

Hydrochloric acid (HCl), nitric acid (HNO_3), and sulfuric acid (H_2SO_4) are known as *mineral acids*. They are some of the strong acids. Acetic acid ($HC_2H_3O_2$), carbonic acid (H_2CO_3), and hydrosulfuric acid (H_2S), which is hydrogen sulfide gas dissolved in water, are some of the weak acids. The strength of an acid depends on the degree of its ionization (ion formation). More ionization means more hydronium ions in the solution and therefore stronger acid.

$$HCl + H_2O \rightarrow H_3O^+ + Cl^-$$

Hydrochloric acid is a strong acid because there is about 99 percent ionization into H_3O^+ and Cl^- ions. Normally, the more hydrogen atoms in the formula, the weaker the acid.

Properties of Acids

1. Acid solutions are sour.
2. Acids form hydronium ions in their water solution.
3. Acids neutralize hydroxides.

   $H_2SO_4 + Ca(OH)_2 \rightarrow CaSO_4 + 2H_2O$
4. Acids react with metals and produce hydrogen gas (H_2).

   $Zn + 2HCl \rightarrow ZnCl_2 + H_2\uparrow$
5. Acids react with oxides of metals to form a salt and water.

   $CuO + H_2SO_4 \rightarrow CuSO_4 + H_2O$
6. Acids react with carbonates to liberate carbon dioxide (CO_2).

   $CaCO_3 + 2HCl \rightarrow CaCl_2 + H_2O + CO_2\uparrow$
7. Acids change the color of certain indicators.

Naming Acids

Binary acids (acids formed of only two elements) start with *hydro-* and end with *-ic*. After hydro, comes the root of the name of the second element.

HCl = hydrochloric acid

HBr = hydrobromic acid

H_2S = hydrosulfuric acid

Ternary acids (acids formed of three elements: hydrogen, middle element, and oxygen) commonly have the root of the name of the middle element with the ending *-ic* or *-ous*.

H_2SO_4 = sulfuric acid

HNO_3 = nitric acid

H_2CO_3 = carbonic acid

H_3PO_4 = phosphoric acid

If the same three elements form more than one acid, use the following system:

$HClO_4$ = per chloric (one oxygen atom more than chloric acid)

$HClO_3$ = chloric acid

$HClO_2$ = chlorous (one oxygen atom less than chloric acid)

$HOCl$ or $HClO$ = hypochlorous (two oxygen atoms less than chloric acid)

Bases

A *base* is a substance that accepts the protons. A common base is the hydroxide ion (OH^-). Because hydroxides such as $Ca(OH)_2$ and $NaOH$ furnish OH^- ions to their solutions, they are commonly known as bases.

$$Ca(OH)_2 \rightarrow Ca^{+2} + 2OH^-$$

$$OH^- + H_3O^+ \rightarrow 2H_2O$$

Some common hydroxides are:

Lye, sodium hydroxide = $NaOH$

Milk of Magnesia, magnesium hydroxide = $Mg(OH)_2$

Lime, calcium hydroxide = $Ca(OH)_2$

Ammonia aqua, ammonium hydroxide = NH_4OH

The strength of a base, like an acid, depends on its degree of dissociation (ion formation). NaOH dissociates strongly and is a strong base; whereas, $Ca(OH)_2$ dissociates poorly and is a weak base.

Properties of Bases

1. Hydroxides are ionic compounds. Their solutions are bitter and slippery.
2. Base solutions change the color of certain indicators.
3. Bases neutralize acids.
4. Bases react with oxides of nonmetals to produce salts and water.

$$CO_2 + Ca(OH)_2 \rightarrow CaCO_3\downarrow + H_2O$$

Amphiprotic Compounds

Amphiprotic compounds behave as proton donors with some substances and proton acceptors with other substances. They accept protons from strong acids and donate protons to strong bases. They are weak bases.

Base $\qquad$ Acid

$$Zn(OH)_2 + 2HCl \rightarrow ZnCl_2 + 2H_2O$$

Acid $\qquad$ Base

$$Zn(OH)_2 + 2NaOH \rightarrow Na_2ZnO_2 + 2H_2O$$

Sodium zincate

Acid		Base		
HCl	+	H_2O	→	$H_3O^+ + Cl^-$

Base		Acid		
NH_3	+	H_2O	→	$NH_4^+ + OH^-$

Zinc hydroxide and water are amphiprotic compounds.

Anhydrides

Anhydrides are oxides that react with water to produce acids or hydroxides. Acid anhydrides are oxides of nonmetals that form acids in water, and basic anhydrides are oxides of metals that form hydroxides in water.

$CO_2 + H_2O$ → H_2CO_3
Acid anhydride — Carbonic acid

$CaO + H_2O$ → $Ca(OH)_2$
Basic anhydride — Calcium hydroxide

Thus, oxides of nonmetals, such as CO_2, cause acidity in water, and soluble oxides of metals, like CaO, cause alkalinity in water.

pH Values

pH or the hydronium ion index scale measures hydronium ion concentration in solutions. Numerically, pH is the common log of the number of litres of a solution containing 1 mol of hydronium ions (19 g) or hydrogen ions (1 g). The number of litres having 1 mol of H_3O^+ ions is the reciprocal of the number of moles of H_3O^+ ions per litre. Therefore, pH is the log of the reciprocal of the H_3O^+ ion concentration expressed as moles per litre (mol/L). Suppose 1 L of a solution contains 0.001 mol of H_3O^+ ions. Then, 1,000 L would contain 1 mol of H_3O^+ ions. And the pH of the solution would be 3 (the common log of 1,000).

$$H_3O^+ \text{ ion concentration} = 0.001 \text{ mol or } 1/1{,}000 \text{ mol/L}$$
$$\text{Reciprocal of } 1/1{,}000 = 1{,}000/1$$
$$= 10^3$$
$$\text{or log of } 1{,}000 = 3$$
$$\text{pH} = 3$$

pH may also be defined as minus log of hydronium ion concentration expressed as moles per litre.

Mathematically, pH = –log $[H_3O^+]$. (The use of brackets [ ] means concentration as moles per litre.)

$\text{pH} = -\log [H_3O^+]$ (In the above example.)

$\text{pH} = -\log 1/1{,}000$

$\text{pH} = -(-3) = 3$

pH Scale

Water at a temperature of 25°C, due to self-ionization, contains 10^{-7} (0.0000001) mol of hydronium ions per litre.

$$H_2O + H_2O \rightleftarrows H_3O^+ + OH^-$$

Therefore, its pH is 7. Hydroxide ion (OH^-) concentration is also 10^{-7} mol/L at the same time. Thus, pH 7 indicates a neutral solution. Water and most of the diluted solutions in a laboratory at 25°C have $[H_3O^+] \times [OH^-] = 10^{-14}$ mol^2/L^2. It is derived by substituting 10^{-7} mol/L for each of the H_3O^+ and OH^- ions. $10^{-7}\text{mol/L} \times 10^{-7}\text{mol/L} = 10^{-14}$ mol^2/L^2, which is known as K_w, the water constant.

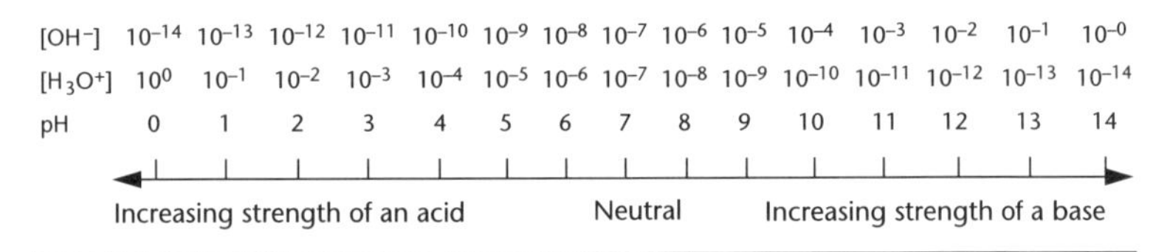

Figure 6-1 pH Scale

The pH scale (Figure 6-1) ranges from 0 to 14, based on the water constant. It measures 1 mol/L of H_3O^+ ion concentration as the maximum at 0 and 1 mol/L of OH^- ion concentration at 14. Therefore, using the previous equation, when H_3O^+ ion concentration is 1 mol/L, OH^- ion is 10^{-14} mol/L, and vice versa.

Methods of pH Measurement

Colorimetric: An indicator such as phenol red or bromthymol blue is added to the solution. It produces a certain intensity of color that is then matched with the permanent standard. pH is estimated to the nearest 0.1 pH unit.

Potentiometer or pH meter: This is the most convenient and accurate method of pH measurement. The meter measures the voltage difference developed between the electrodes due to hydronium ions in the solution. This difference is then read on the scale in pH units.

Hydrion paper: Dip a piece of this paper in the solution and match the color developed with the permanent standard.

Alkalinity and Acidity in Natural Water

Acidity in water is usually due to CO_2 (acid anhydride), mineral acids, or hydrolysis of some heavy metal salts such as aluminum sulfate ($Al_2(SO_4)_3$). Hydrolysis is decomposition of a substance by reacting with water.

$$Al_2(SO_4)_3 + 6H_2O \rightarrow 2Al(OH)_3\downarrow + 3H_2SO_4$$

Alkalinity may be defined as the capacity to neutralize acidity. It is caused in surface waters by hydroxides, carbonates, and bicarbonates.

$$Ca(OH)_2 + H_2SO_4 \rightarrow CaSO_4 + 2H_2O$$

$$CaCO_3 + H_2SO_4 \rightarrow CaSO_4 + Ca(HCO_3)_2$$

$$Ca(HCO_3)_2 + H_2SO_4 \rightarrow CaSO_4 + 2CO_2\uparrow + 2H_2O$$

Salts

A salt is an ionic compound containing anions formed from an acid after the donation of protons. For example, in the case of sodium chloride (NaCl), a salt, Cl^- comes from HCl. In the case of Na_2SO_4, sulfate ($SO_4^=$) comes from H_2SO_4. Some common examples of salts are chlorides, bromides, iodides, nitrates, sulfates, carbonates, and phosphates. A salt is produced when an acid reacts with a base.

$$HCl + NaOH \rightarrow \underset{\text{Salt}}{NaCl} + H_2O$$

Questions

1. Which element is common in all acids?
2. Define acid, base, salt, and pH.
3. A set of solutions has pH values of 3, 4, 7, and 10. Which solution is the most acidic, which one is the most basic, and which one is neutral?
4. What would be the pH of a solution containing 0.001 mol of H_3O^+ ion concentration per litre?
5. Name various methods of measuring the pH of a solution.
6. What is the water constant?
7. Differentiate an acid from a base.
8. Which ion causes acidity?
9. When an acid reacts with a base, what products get produced?
10. Is sodium hydroxide an acid or a base?

7 Solutions

A solution is a homogeneous mixture of two or more substances, with its composition varying within certain limits. A solution, therefore, has uniform composition. Examples are sugar and table salt in water. The dissolving medium is the *solvent* and the dissolved substance, the *solute*. Water in the above-mentioned solutions is the solvent, and sugar and salt are the solutes.

A heterogeneous mixture of two or more substances is called a *suspension*, e.g., clay in water. In this case, suspended particles of the dispersed substance, clay, are distributed nonuniformly in the medium. The bottom part of the mixture has more particles than the upper. When suspended particles are too small to settle by the force of gravity and too large to dissolve to form a solution, the suspension is known as a *colloid*. Particle size in a colloidal suspension is 1–100 nm. A nanometre is equal to 1/1,000,000 mm or 10^{-6} mm. These particles show peculiar dancing movements known as *Brownian motions* due to their continuous collisions with water molecules. Suspended particles of a colloid cause *turbidity*. Colloids are discussed in detail in chapter 8.

Solutions of ionic compounds and acids in water conduct electricity because they produce ions. These substances are known as *electrolytes*. Most of the covalent compounds stay as molecules in their solutions and do not conduct electricity. Such substances are known as *nonelectrolytes*. Sodium chloride and hydrochloric acid are electrolytes, and sugar is an example of a nonelectrolyte.

Solutions can be classified as solids, liquids, and gases, depending on their physical states, which depend on the solvent. Metals mix with metals and form alloys, which are solid solutions; some solids and liquids mix in liquids to form liquid solutions; and gases mix with gases to form gas solutions. Copper in nickel, alcohol in water, and oxygen in nitrogen (air) are examples of solid, liquid, and gas solutions, respectively.

Concentration of a Solution

Percent concentration is parts of the solute in 100 parts of the solution by mass.

Percentage concentration can be expressed as

$$\%\ \text{Concentration} = \frac{\text{Mass of solute}}{\text{Mass of solution}} \times 100$$

Therefore, 10 g of sugar in 90 g of water is a 10 percent sugar solution.

Molarity (M) is the number of moles (gram formula weights of compounds) of the solute per litre of its solution. A molar (*M*) solution contains 1 mol of the solute per litre, and a 2 molar (2*M*) solution contains 2 mol of the solute per litre. A molar solution of HCl, for example, contains about 36.5 g of HCl per litre of solution and is labeled as *M* HCl. A solution with 73 g of HCl per litre is 2*M* HCl and one with 18.25 g of HCl per litre is 0.5*M* HCl. Molarity is also known as *formality*.

M H_2SO_4 contains 98 g of H_2SO_4 per litre of the solution.

0.25*M* H_2SO_4 contains 98 × 0.25 or 24.5 g of H_2SO_4 per litre of the solution.

M NaOH contains 40 g/L of the solution.

0.10*M* NaOH contains 40 × 0.1 or 4 g/L of the solution.

Molality (m) is the number of moles of the solute per kilogram of the solvent. It expresses the quantity of solute in quantity of the solvent. One mol of a solute dissolved in 1 kg of the solvent forms a "molal" solution.

Normality (N) is the number of gram equivalent weights of a solute per litre of the solution. Gram equivalent weight of a substance may be defined as the mass of a substance in grams that contains, replaces, or reacts with the Avogadro number of hydrogen atoms (a mole or 1 g of hydrogen atoms). A mole of hydrogen atoms offers an Avogadro number of electrons in a reaction because a hydrogen atom has only one electron to offer. Therefore, gram equivalent weight may be defined as the mass of a substance in grams that has an Avogadro number of electrons to participate in a chemical reaction. Gram equivalent weights of various substances are calculated by the following formulae:

$$\text{Gram equivalent weight of an element} = \frac{\text{mole or gram atomic weight}}{\text{oxidation number}}$$

Examples:

$$\text{Gram equivalent weight of hydrogen} = \frac{1\text{ g}}{1} = 1\text{ g}$$

$$\text{Gram equivalent weight of chlorine} = \frac{35.5\text{ g}}{1} = 35.5\text{ g}$$

$$\text{Gram equivalent weight of oxygen} = \frac{16\text{ g}}{2} = 8\text{ g}$$

$$\text{Gram equivalent weight of a compound} = \frac{\text{mole or gram formula weight}}{\text{positive charge in the formula}}$$

Examples:

$$\text{Gram equivalent weight of HCl} = \frac{36.5\text{ g}}{1} = 36.5\text{ g}$$

Gram equivalent weight of $H_2SO_4 = \frac{98\ g}{2} = 49\ g$

Gram equivalent weight of H_3PO_4 (phosphoric acid) =

$\frac{98\ g}{3} = 32.66\ g$

Gram equivalent weight of $Ca(OH)_2 = \frac{74\ g}{2} = 37\ g$

Gram equivalent weight of $Al_2(SO_4)_3 = \frac{342\ g}{6} = 57\ g$

Gram equivalent weight of $CaCO_3 = \frac{100\ g}{2} = 50\ g$

In case of acids and hydroxides, gram equivalent weight may be determined by dividing the mole by the replaceable hydrogen atoms and hydroxide ions, respectively.

In case of redox reactions, the gram equivalent weight of a reactant is calculated by dividing its mole by the total change in its oxidation number in its one formula unit. The total change in oxidation number is determined by determining oxidation numbers of all the elements in the formula on the reactant and product sides of the equation. Study the following examples.

Examples:

Determine gram equivalent weights of potassium permanganate ($KMnO_4$), potassium dichromate ($K_2Cr_2O_7$), and sodium thiosulfate ($Na_2S_2O_3$) in the following redox reactions:

a.

$$\overset{+1\ +7-2}{2K\,MnO_4} + 5H_2C_2O_4 + 3H_2SO_4 \rightarrow \overset{+1}{K_2SO_4} + \overset{+2}{2MnSO_4} + 10CO_2 + \overset{-2}{8H_2O}$$

Total oxidation numbers $\quad (+1\ +7\ -8) \quad (+1\ +2\ -8)$

$\qquad K\ Mn\ O_4 \rightarrow K, Mn, 4O$

Difference = +5

Gram equivalent weight of $KMnO_4$ in this reaction $= \dfrac{\text{mole}}{5}$

$= \dfrac{158}{5}$ g

= 31.6 g

b.

$$\overset{+1\ +6\ -2}{K_2Cr_2O_7} + 6KI + 7H_2SO_4 \rightarrow \overset{+1}{4K_2SO_4} + \overset{+3}{Cr_2(SO_4)_3} + \overset{-2}{7H_2O} + 3I_2$$

Total oxidation numbers $\quad (+2\ +12\ -14) \quad (+2\ +6\ -14)$

$\qquad K_2Cr_2O_7 \rightarrow 2K, 2Cr, 7O$

Difference = +6

Gram equivalent weight of $K_2Cr_2O_7 = \dfrac{\text{mole}}{6}$

$= \dfrac{294}{6}$ g

= 49 g

c.

$$\overset{+1\ +2\ -2}{2\,Na_2\,S_2O_3} + I_2 \rightarrow \overset{+1\ +2.5-2}{Na_2S_4O_6} + 2NaI$$

Total oxidation numbers $\quad (+2\ +4\ -6) \quad (+2\ \ +5\ \ -6)$

$\qquad Na_2S_2O_3 \rightarrow 2Na, 2S, 3O$

Difference = –1

Gram equivalent weight of $Na_2S_2O_3 = \dfrac{\text{mole}}{1}$

$= \dfrac{158}{1}$ g

= 158 g

The last two chemicals are used in chemical oxygen demand and dissolved oxygen determinations and their gram equivalent weights used are 49 g and 158 g, respectively.

A normal solution contains one gram equivalent weight of a solute per litre, and a 0.5*N* solution contains 0.5 g equivalent weight per litre of the solution.

N HCl contains 36.5 g of HCl per litre of the solution.

0.5*N* HCl contains 18.25 g of HCl per litre of the solution.

N H_2SO_4 contains 49 g of H_2SO_4 per litre of the solution.

0.02*N* H_2SO_4 contains 0.98 g of H_2SO_4 per litre of the solution.

Saturated Solution

A saturated solution contains maximum dissolved solute under specified conditions. Dissolved particles of the solute are in equilibrium with the undissolved particles. The rate of dissolving is equal to the rate of crystallizing.

$$NaNO_3(s) \underset{\text{Crystallizing}}{\overset{\text{Dissolving}}{\rightleftarrows}} Na^+(aq) + NO_3^-(aq)$$

where:

s = solid
aq = dissolved

Unsaturated Solutions

A solution with any amount of dissolved solute less than the amount required to make a saturated solution is considered to be unsaturated.

Supersaturated Solutions

A supersaturated solution contains more dissolved solute than required for preparing a saturated solution and can be prepared by heating a saturated solution, adding more solute, and then cooling it gently. Excess dissolved solute crystallizes by seeding supersaturated solution with a few crystals of the solute.

Dilute and Concentrated Solutions

These terms are vague as they lack quantitative preciseness. A dilute solution contains less solute and a concentrated solution, more.

Standard Solutions

A standard solution is an accurately prepared solution used to determine the concentration of other solutions, which are then known as *standardized solutions*. In the laboratory, we can use a standard acid solution to standardize a base solution, and vice versa. The volumetric technique of matching a standard solution and a solution of unknown concentration is known as *titration*. This process of comparison is known as *standardization*. To indicate the completion of a reaction, another substance called an *indicator* is used. An indicator has a different color in the standard and the standardized solutions. For example, let's standardize a solution of sodium hydroxide (NaOH) with a standard sulfuric acid (H_2SO_4) solution. Add a few drops of phenolphthalein, the indicator, to a known volume of NaOH solution and the solution will turn pink. Then slowly add a standard H_2SO_4 solution to it until the pink color disappears. Phenolphthalein is colorless in acid solutions.

Determine concentration (normality) of NaOH solution by using the equation

$$NV = N_1V_1$$

where:

N = normality of standard
V = volume (mL) of standard
N_1 = normality of the standardized
V_1 = volume of the standardized

Suppose in the above titration we took 100 mL of NaOH solution and used 50 mL of 0.1*N* H_2SO_4 to reach the end of the reaction or equivalence point (end point). Normality of NaOH can be calculated as

$$\frac{H_2SO_4}{NV} = \frac{NaOH}{N_1V_1}$$

$$0.1 \times 50 = N_1 \times 100$$

$$N_1 = \frac{0.1 \times 50}{100} = 0.05$$

Sodium hydroxide is 0.05*N* NaOH solution. This equation can also be used for making dilute solutions from stock solutions.

Suppose we prepare 1,000 mL of 0.02*N* H_2SO_4 from 1*N* H_2SO_4 stock solution. It requires only 20 mL of 1*N* H_2SO_4 to be diluted in 1 L according to the equation.

$$\frac{\text{Stock}}{NV} = \frac{\text{Dilute}}{N_1V_1}$$

$$1 \times V = 0.02 \times 1{,}000 \text{ mL}$$

$$V = 20 \text{ mL}$$

This equation is based on the fact that the volume of a solution is inversely proportional to the concentration and also that equal volumes of solutions with the same normality are chemically equivalent.

Factors That Affect Solubility

Temperature: A rise in temperature usually increases solubility of a solid in a liquid except for a few substances such as slaked lime ($Ca(OH)_2$) and sodium sulfate (Na_2SO_4). In the case of gases, however, the lower the temperature, the higher the solubility, and vice versa. For example, dissolved oxygen in surface waters is 14.62 and 9.17 mg/L at 0° and 20°C, respectively.

Pressure: Pressure has an insignificant effect on the solubility of a solid or liquid in a solid or liquid. Solubility of a gas in a liquid or a solid increases with the rise in pressure, and vice versa. Henry's law states that the mass of a dissolved gas in a liquid is directly proportional to the pressure of the gas in contact with the liquid, at a constant temperature.

Carbonated beverages in a capped bottle have 5–10 atmospheres pressure. Without the cap, there is effervescence (rapid evolution of a gas from a solution) due to lower solubility at a lower pressure.

Nature of the solute and the solvent: Like substances dissolve in like solvents. Inorganic substances dissolve in inorganic solvents and organic substances, in organic solvents. Solubility of a substance depends on two types of forces: attractive forces among the solute particles and attractive forces between solute and solvent particles. The former cause lower solubility and the latter, more.

Questions

1. Define the following:
   a. Solution
   b. Solute
   c. Solvent
   d. Colloidal suspension
   e. Gram equivalent weight

f. Saturated solution

g. Unsaturated solution

h. Standard solution

2. Calculate gram equivalent weights of the following:

a. HCl

b. H_2SO_4

c. NaOH

d. Na_2CO_3

e. $Fe_2(SO_4)_3$

3. How many grams of NaOH are required to prepare 500 mL of 0.1*N* NaOH solution?

4. Calculate the normality of a HCl solution in a titration when 100 mL of it used 75 mL of 0.25*N* NaOH. What will be its molarity?

5. a. State Henry's law.

   b. Will there be more or less dissolved oxygen at (1) higher oxygen pressure and (2) at higher temperature?

6. Differentiate a solution from a suspension.

7. How many different states are in solutions?

8. What is meant by molarity of a solution?

9. Differentiate between a normal and a molar solution.

8 Colloids and Coagulation

The term *colloid* means *glue* in Greek and it was coined by Thomas Graham in 1861. A colloid is formed of dispersed particles and dispersing medium. The dispersed particles range from 1–100 nm in diameter as compared to particles of true solutions, which are 0.1–1 nm (see Table 8-1).

Colloidal particles may be aggregates of atoms or molecules that are big enough to be filtered out and small enough to stay suspended against the force of gravity. These particles are visible with a light-powered microscope. Some common examples of colloids are fog, smoke, foam, cream, cheese, paints, and glue.

Suspension			
Coarse	Fine	Colloid	Solution
1,000	100–1,000	1–100	0.1–1

Table 8-1 Particle Size of Suspensions and Solutions (nm)

Methods of Colloid Formation

Any insoluble material in the dispersion medium can produce a colloid. Colloidal particles can be produced by the following:

1. Colloid mills . . . coarse grinders.
2. Ionic reactions . . . formation of insoluble products.
3. Mixing of some insoluble organic compounds in water, such as soap, starch, gelatin, agar, gum arabic, and albumin. In these cases, water can disperse the particles only into colloidal suspension.

Types of Colloids

Classification of colloids is based on the state of the dispersed phase and the dispersion medium. There are only eight classes because a gas does not form suspension in a gas.

Colloidal Dispersion	Examples
Solid in solid	Ruby glass
Solid in liquid	Glue, India ink
Solid in gas	Smoke
Liquid in solid	Jelly
Liquid in liquid	Cream, emulsions
Liquid in gas	Fog
Gas in solid	Floating soap
Gas in liquid	Foam, whipped cream

Colloids may be classified into sols or gels based on their liquid or jelly-like state, respectively. Or, colloids may be classified into lyophilic or lyophobic colloids, depending on the affinity of the dispersed phase with the dispersion medium.

1. *Lyophilic (medium-loving) colloids:* If a small amount of powdered gelatin is mixed in water and dispersion is allowed to stand, it sets to form a gel or jelly. The dispersed particles of gelatin have great affinity for the dispersing medium and become thoroughly hydrated. This traps the water in such a way that viscosity of dispersion increases, which is known as *emulsoid*. Other examples are soap, soluble starch, soluble protein, blood serum, and gum arabic.

2. *Lyophobic (medium-hating) colloids:* Dispersed particles have a poor affinity for the dispersing medium and, thus, there is no significant change in the viscosity of the medium. They are, therefore, as fluid as the medium and are highly susceptible to coagulation, which is the separation of colloidal particles from the medium by adding electrolytes, e.g., coagulants like alum and ferric sulfate in water treatment. These colloids are known as suspensoids. The terms *hydrophilic* and *hydrophobic* are used when the dispersion medium is water.

Colloids can also be classified as positive and negative (see Figure 8-1). All colloids are electrically charged, but the charge varies considerably in magnitude with the nature of the colloidal material. A colloidal charge may change with a change in external conditions. For this reason, both $Al(OH)_3$ and $Fe(OH)_3$ flocs may be positive or negative. The charge is gained by the selective adsorption of ions from the surrounding medium or due to ionization of surface components of the colloid particles. Most colloids in water that cause turbidity and color are negative.

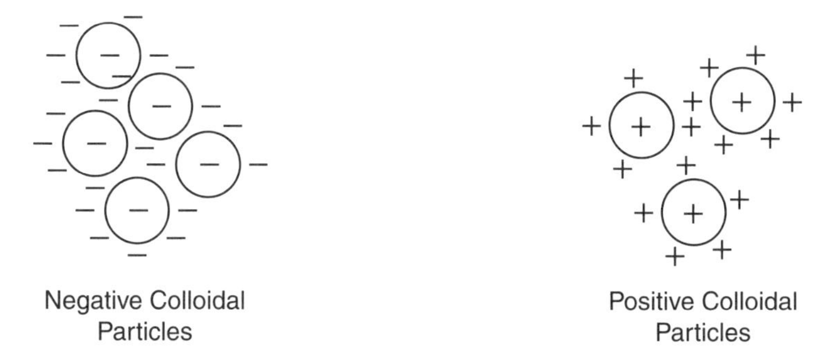

Figure 8-1 Negative and Positive Colloidal Particles

General Properties

Colloidal particles are very small, but their surface area in relation to mass is very great. If a 1-cm^3 cube is divided into a 1-nm^3 cubes, the surface area increases from 6 cm^2 to 600 m^2. Due to this fact, surface phenomena predominate and control the behavior of colloids. The mass of colloidal particles is so small that gravitational effects are unimportant.

Adsorption

Adsorption is the acquiring of a gas, liquid, or solid on the surface of a liquid or solid with which it is in contact. Adsorbing material is known as *adsorbent* and adsorbed as *adsorbate*. For example, powdered activated carbon is used in water treatment as adsorbent for pesticides, and chemicals causing tastes and odors are the adsorbates. Because of the large surface area of dispersed particles of a colloid, a remarkable amount of adsorption may occur. Adsorption is generally preferential in nature, with some ions preferred and others excluded. This selective action yields a charge on colloidal particles and is important in the stability of many colloidal dispersions.

Effect on Freezing and Boiling Points

Colloidal dispersions have less effect on the freezing and boiling points of liquids than the solutes due to the fact that dispersions are very diluted with only a few dispersed particles. Each particle may be formed of hundreds or thousands of molecules.

Dialysis

The separation of colloids from crystalloids by diffusion through a semipermeable membrane is dialysis. Due to their large particle size, colloids do not pass through a semipermeable membrane, whereas solute particles do.

Electrical Properties

Colloidal particles are charged, as mentioned before, due to the adsorption of ions or ionization on the surface of the colloidal particle. Many colloidal dispersions are dependent on the electrical charge for their stability. Like charges repel; therefore, similarly charged particles of colloid do not form aggregates or floc, thus making the system stable. The magnitude of the charge at the boundary between the volume held (total volume of colloidal particles, adsorbed ions, and water associated with the aggregate) and the surrounding medium is known as *zeta potential*. The magnitude of zeta potential may be estimated from electrophoretic measurements of particle movements.

Zeta potential or repulsion between colloidal particles is counteracted by van der Waals forces, Brownian motions, and gravitational forces to destabilize or coagulate the system. The *van der Waals forces* are intermolecular attractive forces. *Brownian motions* are random movements of colloidal particles caused by bombardment of molecules of the dispersion medium. Brownian motions, when vigorous due to high temperature, cause effective collisions among colloidal particles. The objective of chemical coagulation is to reduce the zeta potential to almost zero for effective turbidity removal.

Tyndall Effect

Because colloidal particles have dimensions greater than the average wavelength of white light, they reflect light. Therefore, a beam of light, passing through a colloidal suspension, is visible to an observer at the right angle of the beam.

Methods of Precipitating Colloids

Electrophoresis

This is the coagulation of a colloid in an electrical field. Colloid particles will accumulate at the pole of opposite charge.

Addition of a Colloid of Opposite Charge

When colloids of opposite charge are mixed, there is mutual precipitation. This method is used less frequently in sanitary engineering because large volumes of a second colloid will be required.

Addition of Electrolytes

Electrolytes contribute ions of opposite charge than that of a colloid. In case of water and wastewater treatment, the more the charge on the cation, the better and more effective the coagulation. This phenomenon was worked out by Schultze and Hardy. The *Schultze–Hardy rule* states that the precipitation of a colloid is affected by that ion of an added electrolyte, which has a charge opposite in sign to that of the colloidal particles, and the effect of such ion increases markedly with the number of charges it carries. The precipitation effect of a bivalent ion is 50–60 times as great as a monovalent ion, and that of a trivalent ion is 600–700 times as great as a monovalent ion. This is the reason that aluminum and iron compounds are used as common coagulants in water and sewage treatment.

Cations or ions of opposite charge force their way into the colloidal particles and, thus, decrease the zeta potential and form the floc.

Trivalent salts in coagulation of water act in three capacities: (a) Salts ionize to yield trivalent cations, some of which reach targets and neutralize colloidal particles. (b) Others combine with hydroxide ions and form colloid metallic hydroxides. (c) These hydroxides precipitate out, being insoluble and in excess of the amount needed to neutralize colloids causing color and turbidity. Sulfates neutralize the positive colloidal metallic hydroxides more effectively than chlorides. That is why they are preferred in water and sewage treatment.

Coagulation occurs in three steps. First, by rapid mixing of Al^{+3} or Fe^{+3} ions, a considerable portion of the colloids is neutralized. Second, due to aggregation, they form microfloc, which is still invisible. Microfloc, due to its positive charge, still attracts negative ions. At the third stage, microfloc particles agglomerate with the formation of large floc particles that can settle.

Boiling

Boiling of a hydrophobic colloidal dispersion often results in coagulation. This is due to the modification in the degree of hydration and higher Brownian motions, which bring about agglomeration.

Freezing

During the freezing process, crystals of pure water form, thus, the concentration of colloids and crystalloids increases and coagulation occurs.

Questions

1. Define the following:
   a. Colloid
   b. Lyophilic
   c. Zeta potential
2. Give three general properties of colloids.
3. Give common methods of precipitating colloids.
4. Explain the Schultze–Hardy rule.
5. Name various coagulants used in water and wastewater treatment.

9 Water

Water is the most abundant and most useful liquid in nature. Surface water covers about three-fourths of the earth's surface. Furthermore, there is subterranean water known as groundwater. Water is formed of 11.11 percent hydrogen and 88.89 percent oxygen. Sixty-five percent of our body weight is water.

Physical Properties of Water

Pure water is a colorless, odorless, and tasteless liquid. The blue or bluish green tint of water is due to its depth. Various dissolved substances, such as minerals and gases like sulfur dioxide and chlorine, may cause tastes and odors. Water's freezing point is 0°C and its boiling point is 100°C. When frozen, it expands by one-ninth of its original volume. Therefore, the density of ice is 0.9 g/cm^3. This expansion causes bursting of some pipes during winter. Water has a maximum density of $1g/cm^3$ at 4°C and, therefore, is lighter above and below 4°C. This variation of density at different temperatures causes the stratification of lakes.

The boiling point of water depends upon the pressure on its surface. The higher the pressure, the higher the boiling point. At 15 psi pressure, water boils at 121°C. This principle is used by a pressure cooker for quick cooking and sterilization. By lowering the pressure, the boiling point is lowered for quick evaporation of water. This principle is used for condensing milk, etc. One litre of water forms 1,700 L of steam at normal atmospheric pressure.

Structure and Properties of Water Molecules

A water molecule is formed of two atoms of hydrogen and one atom of oxygen bonded by polar covalent bonds. The angle between the two bonds is 105°. Since oxygen is more strongly electronegative than hydrogen, there is a slight negative charge at the oxygen part and a slight positive charge at the hydrogen parts. This polarity causes water molecules to join together to form groups of 4–8 molecules depending upon the temperature, where the hydrogen part of one molecule is weakly bonded with the oxygen part of the other molecule, and vice versa. This weak bonding is known as *hydrogen bonding*. This association of water molecules results in the higher boiling point of water as compared to other covalent compounds of similar molecular weight. It also causes an open hexagonal grouping in the ice, which makes the water expand when frozen. At 4°C, there is collapsing of the hexagonal pattern and the maximum crowding of water molecules, which results in its maximum density. Above 4°C, the higher kinetic energy pushes water molecules further apart and the density decreases. At the boiling point, all hydrogen bonds are broken and water molecules are free to vaporize.

Water as a Standard

Due to water's unique physical properties and its availability in a pure state, water is used as a standard for several parameters.

1. *Gram:* Gram is mass of 1 cm^3 of water at 4°C.
2. *Temperature scale:* Under the standard atmospheric pressure, the freezing point of water is 0°C and the boiling point is 100°C.
3. *Calorie:* It is the unit of heat. Calorie is the amount of heat required to raise the temperature of 1 g of water by 1°C.

4. *Specific gravity:* Density of water is used to determine the specific gravity of solids and liquids.

Chemical Behavior

Water is a universal solvent and it reacts with and dissolves a large number of substances, including the following:

1. *Metals:* A number of metals react with water to form hydroxides and hydrogen gas.

$$2Na + 2H_2O \rightarrow 2NaOH + H_2\uparrow$$

$$Ca + 2H_2O \rightarrow Ca(OH)_2 + H_2\uparrow$$

2. *Metallic oxides:* Water reacts with the soluble oxides of metals, such as sodium, potassium, calcium, and aluminum, to form hydroxides.

$$CaO + H_2O \rightarrow Ca(OH)_2$$

$$Al_2O_3 + 3H_2O \rightarrow 2Al(OH)_3\downarrow$$

These oxides of metals are known as basic anhydrides.

3. *Oxides of nonmetals:* Oxides of nonmetals react with water to form acids.

$$CO_2 + H_2O \rightarrow H_2CO_3$$

These oxides are known as acid anhydrides. Other examples of acid anhydrides are sulfur oxides, nitrogen oxides, and phosphorus oxides.

4. *Water of crystallization:* When water is evaporated from a solution of certain crystalline compounds, a specific number of water molecules stays associated with cations and anions to maintain the crystalline state. This specific number of water molecules is known as *water of crystallization* or *hydration*. These crystalline compounds are called *hydrates*. Water of

crystallization is written after the formula with a raised dot indicating the loose attachment.

For example:

$$\text{Alum} = Al_2(SO_4)_3 \cdot 14H_2O$$

$$\text{Copper sulfate} = CuSO_4 \cdot 5H_2O$$

$$\text{Ferrous sulfate} = FeSO_4 \cdot 7H_2O$$

When a hydrate is heated above 100°C, the water is evaporated, the crystals crumble, and the compound becomes a dry powder known as *anhydrous*.

Some hydrates have a higher vapor pressure and lose water of crystallization when exposed to the atmosphere. This phenomenon is known as *efflorescence*. For example, $Na_2CO_3 \cdot 10H_2O$ (sodium carbonate dekahydrate) becomes $Na_2CO_3 \cdot 1H_2O$ (monohydrate).

The opposite of efflorescence is *deliquescence*. Certain hydrates, due to their lower water vapor pressure than that of the atmosphere, will gain water when exposed to air, e.g., calcium chloride ($CaCl_2$) and magnesium chloride ($MgCl_2$), which is a very deliquescent compound. These substances are used in desiccators. Any substance that gains moisture is known as *hygroscopic*.

Questions

1. Give four important properties of water.
2. Why is ice lighter than water?
3. At what temperature does water have maximum density and why?
4. Define the following terms:
   a. Water of crystallization
   b. Deliquescence
   c. Efflorescence

5. Why does water have a higher boiling point than other covalent compounds with about the same molecular weight?
6. Complete and balance the following equations:
   a. $SO_2 + H_2O \rightarrow$
   b. $Na + H_2O \rightarrow$
   c. $CaO + H_2O \rightarrow$

10 Ionization Theory

Ionization theory deals with ion formation by electrolytes and the behavior of their solutions. Ionic compounds and acids conduct electricity in their water solutions and are known as *electrolytes;* whereas, covalent compounds (other than acids) do not conduct electricity in their water solutions and are called *nonelectrolytes*. Because electrolytes exist as ions (their cations move toward the negative pole and anions, toward the positive pole), electricity is conducted. Nonelectrolytes do not form ions in their solutions and they do not conduct electricity.

Some common examples of electrolytes are sodium chloride (NaCl) and sulfuric acid (H_2SO_4), while nonelectrolytes include sugars and various other organic compounds.

Michael Faraday (1791–1867), an English chemist, first observed that conducting solutions have charged particles. He named these charged particles *ions* (wanderer) and thought they were produced by electrodes. He also coined the terms *electrolytes* and *nonelectrolytes*.

The Swedish scientist Svante Arrhenius published a report, in 1887, on the theory of ionization. He considered ions to be electrically charged and formed by ionization of molecules.

In modern theory, the solvent has an important role in the solution process. The polar covalent nature of water molecules is very important in understanding ion formation.

Assumptions of Modern Theory

1. An electrolyte in solution forms ions.
2. An ion is an atom or a radical that carries a charge.
3. A water solution of an electrolyte carries an equal number of positive and negative charges.

Dissociation of Ionic Compounds

In forming ions, atoms become electrically charged to gain chemical stability. Ionic compounds exist as crystals made up in an orderly manner, with every anion close to a cation.

When some crystals of sodium chloride (NaCl) are dropped into a beaker of water, water dipoles exert an attractive force on the ions of the surface layer of the crystal. The negative oxygen poles of a number of water molecules associate with the sodium cations (Na^+). Similarly, the positive hydrogen part of water molecules is attracted by the chloride (Cl^-) anions. The association of water molecules with the ions is known as *hydration of ions*. These ions are then separated from the crystals and diffuse throughout the solution, loosely bonded to the water molecules.

Sodium chloride is said to be dissociated when it is dissolved in water (see Figure 10-1). The symbol for solid is "s"; for liquid, "l"; and for gas, "g." We can represent dissociation of sodium chloride (NaCl) as

$$NaCl\ (s) \rightarrow Na^+\ (aq) + Cl^-\ (aq)$$

The number of water molecules needed for dissociation of an ion depends upon the size and charge of the ion. The more the charge and the heavier the ion, the greater the number of water molecules needed for its dissociation. Water molecules are interchanged from ion to ion and from ion to solvent. In certain cases, water molecules are not involved in reforming the crystal structure

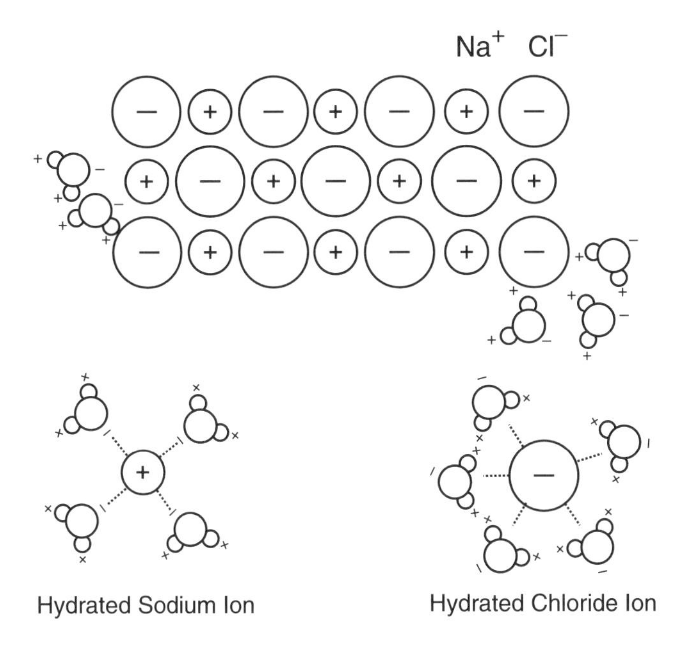

Figure 10-1 Dissociation of Sodium Chloride

after the evaporation of solvent (e.g., sodium chloride (NaCl)). However, in case of hydrates, a specific number of water molecules is retained by the ions in forming the crystal, which is known as *water of crystallization* (discussed in chapter 9). Extensive hydration of the solute ions ties up a large number of water molecules. This reduces the number of free water molecules in the spaces separating hydrated ions of opposite charge. The attraction between ions of opposite charge becomes stronger and crystals begin to form again. The saturation point is reached as the rate of dissolving equals the rate of crystallizing.

Some ionic compounds are only slightly soluble because a large number of water molecules is required to separate their ions, e.g., silver chloride (AgCl), calcium carbonate ($CaCO_3$), magnesium hydroxide ($Mg(OH)_2$), and aluminum hydroxide ($Al(OH)_3$). These are known as *insoluble compounds.*

Both silver nitrate ($AgNO_3$) and sodium chloride (NaCl) are soluble in water, whereas silver chloride (AgCl) is only slightly soluble. Ag^+ (aq) + Cl^- (aq) ions will be present only in a very small number in a saturated solution of AgCl.

If concentrated solutions of NaCl and $AgNO_3$ are mixed, there will be four types of ions in a single solution environment: Na^+, Cl^-, Ag^+, and NO_3^-.

$$NaCl\ (s) \longrightarrow Na^+\ (aq) + Cl^-\ (aq)$$

$$AgNO_3\ (s) \longrightarrow Ag^+\ (aq) + NO_3^-\ (aq)$$

Ag^+ (aq) + Cl^- (aq) = precipitate

The concentration of Ag^+ (aq) and Cl^- (aq) will greatly exceed the solubility of AgCl. All Ag^+ and Cl^- in excess of a saturated solution will separate from the solution as a precipitate of a solid, AgCl. The separation of a solid from a solution is called *precipitation.* As $Na^+ + NO_3^-$ did not take any part in the reaction, they are known as *spectator ions.* This reaction can be shown as a simplified equation:

$$NaCl + AgNO_3 \rightarrow AgCl\downarrow + NaNO_3$$

This is the basis of ionic or double-replacement reactions. This principle is used in the flocculation and softening of water.

$$Al_2(SO_4)_3 + 3Ca(OH)_2 \rightarrow 2Al(OH)_3\downarrow + 3CaSO_4$$

$$CaSO_4 + Na_2CO_3 \rightarrow CaCO_3\downarrow + Na_2SO_4$$

$$MgSO_4 + Ca(OH)_2 \rightarrow Mg(OH)_2\downarrow + CaSO_4$$

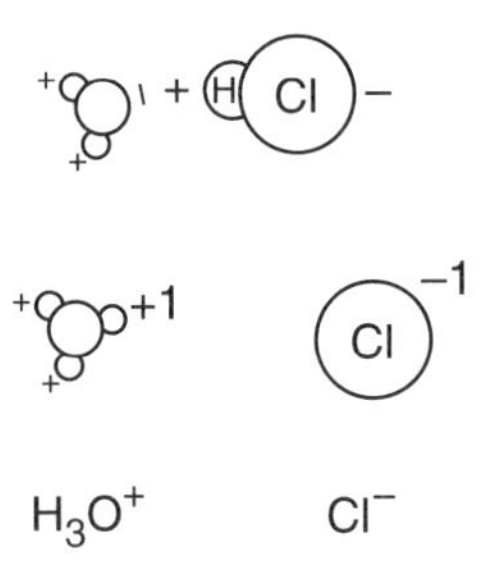

Figure 10-2 Ionization of Hydrochloric Acid

Ionization of Polar Covalent Compounds (Acids)

Acids are polar covalent compounds having the hydrogen part of the molecule with a positive pole and the other part with a negative pole.

The covalent bond is generally a strong bond. However, when acid molecules are in water, the association of water molecules weakens the bond enough to pull the molecule apart. The hydrogen part separates as hydrogen ion (H^+), which associates with a water molecule to form a hydronium ion (H_3O^+), while the rest of the acid molecule forms an anion.

The ions did not exist in the undissolved acid but were formed by the action of water. This process of ion formation from the polar covalent solutes by the action of water molecules is known as *ionization* (see Figure 10-2).

Acids that ionize completely in water are known as *strong acids* and those that ionize poorly are called *weak acids*.

Acetic acid:

$$\underset{99\%}{H\,C_2H_3O_2 + H_2O} \rightleftarrows \underset{1\%}{H_3O^+ + C_2H_3O_2^-} \quad \text{weak}$$

Hydrochloric acid:

$$\underset{1\%}{HCl + H_2O} \rightleftarrows \underset{99\%}{H_3O^+ + Cl^-} \quad \text{strong}$$

Strong and weak refer to the degree of ionization, and dilute and concentrated refer to the amount of acid dissolved in the solvent.

Ionization of Water

Water, being a polar covalent compound, ionizes very poorly to form hydronium (H_3O^+) and hydroxide (OH^-) ions. This low concentration of ions, however, is insignificant when compared with electrolytes such as NaCl and H_2SO_4. Therefore, for all practical purposes, pure water is considered a nonelectrolyte.

$$H_2O + H_2O \rightleftarrows H_3O^+ + OH^-$$

Nonelectrolyte covalent compounds that do not ionize have very little (0 to 0.4) electronegativity difference between combined atoms and, thus, they are not affected by water molecules to form ions.

Effects of Electrolytes on the Freezing and Boiling Points of the Solvent

The freezing and boiling points depend on the number of particles in the solution. The higher the number of particles, the lower the freezing point and the higher the boiling point. An electrolyte ionizes and contributes two, three, or more times the number of particles than a nonelectrolyte. Electrolytes, therefore, are correspondingly more effective than nonelectrolytes in lowering the freezing point and raising the boiling point of the solvent. One Avogadro number of solute particles per kilogram of water raises the boiling point by about 0.52°C and lowers the freezing point by 1.86°C.

Electrolysis and Ionization

Electrolysis is the decomposition of a substance with electricity. Because water ionizes only slightly in the decomposition of water, additional ions must be provided for an adequate conduction of electricity between the electrodes. Sulfuric acid (H_2SO_4) is used for this purpose. Any substance that forms ions can undergo electrolysis, e.g., water and ionic compounds.

In the electrolysis of water when electrodes are connected to a source of direct current, OH^- and $SO_4^=$ ions start going to the anode (positive pole) and H_3O^+ to the cathode (negative pole).

H_3O^+ gains an electron and combines with another H_3O^+ ion to form H_2O and H_2; $4OH^-$ ions lose an electron each and form two molecules of H_2O and O_2.

$$2H_3{}^+O + 2e^- \rightarrow 2H_2O + H_2\uparrow \text{ at cathode}$$

$$4OH^- - 4e^- \rightarrow 2H_2O + O_2\uparrow \text{ at anode}$$

O_2 is liberated at the anode in preference to other nonmetallic ions and H_2 in preference to metallic ions at the cathode.

Electrolysis of sodium chloride (NaCl) yields sodium at the cathode and chlorine at the anode (Figure 10-3).

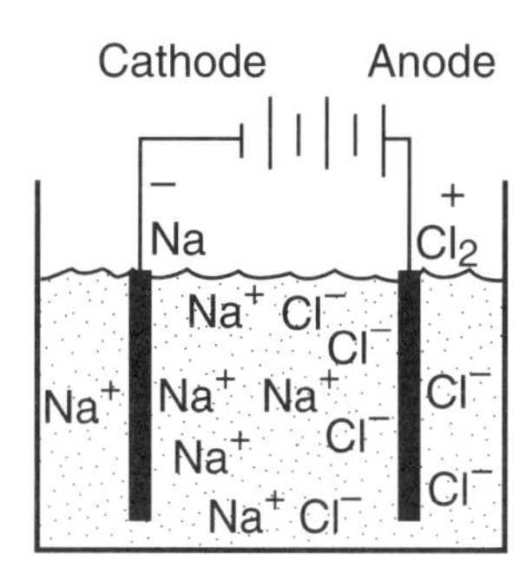

Figure 10-3 Electrolysis of Sodium Chloride

Questions

1. Define the following:
   a. Electrolyte
   b. Nonelectrolyte
   c. Ionization
   d. Dissociation
   e. Precipitation
2. What is the effect of the dissolved electrolytes and nonelectrolytes on the boiling point and the freezing point of water?
3. Give three assumptions of the ionization theory.

11 The Gas Laws

Gas is the lightest, thinnest, and most fluid form of matter. Gases pass through each other more easily than do other forms of matter.

Kinetic Theory

The particles composing all matter are atoms, molecules, or ions. Kinetic theory helps explain the properties of gases, liquids, and solids in terms of the forces between the particles of matter and the energy they possess.

There are three basic assumptions of the kinetic theory:

1. Matter is composed of very tiny particles. The chemical properties of these particles depend on their composition. Their physical properties depend on the forces they exert on each other and the distance separating them.
2. These particles of matter are in constant motion. Their average energy of motion depends on temperature. The higher the temperature, the greater the energy.
3. The particles of matter do not lose energy in collisions. Collisions with each other or with walls of a container are perfectly elastic.

We can explain the behavior or properties of gases using the kinetic theory.

Properties of Gases

1. *Expansion:* Gases do not have a definite volume or a definite shape. All gases are made of millions of tiny particles that move freely at random. As these particles are widely separated, they travel in random directions at a high speed of about 10^3 m/sec. They undergo about 5.0×10^9 collisions per second. Matter in a gaseous state occupies about 1,000 times more volume than in a solid or liquid state. One mL of gas contains about 3.0×10^{19} molecules. An ordinary molecule is about 4 Å in diameter. They are about 40 Å apart.
2. *Pressure:* Gases exert pressure. Pressure is due to the bombardment of molecules with the walls of the container. If we put more gas molecules in the same space, there will be more bombardment and, thus, a higher pressure. If the temperature is raised, more kinetic energy causes higher pressure.
3. *Low density:* The density of a gas is very low. It is about 10^{-3} times the density of the same substance in a liquid or solid state.
4. *Diffusion:* The spontaneous scattering of the gas molecules to fill a space uniformly is diffusion. Gases, therefore, do not have a definite volume or shape. If the stopper of an ammonia bottle is removed, its irritating smell is evident throughout the room rather quickly. This is due to the diffusion of ammonia molecules, which are spaced well apart and are independently moving in all directions.
5. *Plasma:* A gas at a very high temperature or by passing an electric current becomes ionized into electrons and protons and is known as plasma, the fourth state of matter.

Henry's Law

The mass of a dissolved gas in a liquid is directly proportional to the pressure of the gas in contact with the liquid at a constant temperature. Due to lower atmospheric pressure at an elevation of 9,642 ft, solubility of oxygen at 20°C in the surface water is 6 mg/L. At sea level at the same temperature, it is 9.17 mg/L.

Mathematically,

$$S_1P = SP_1 \text{ or}$$

$$\frac{S_1}{P_1} = \frac{S}{P}$$

where:

S = solubility at pressure P
S_1 = solubility at pressure P_1

Suppose the solubility of a gas at 1 atmospheric pressure is 5 mg/L, its solubility at 3 atmospheric pressure will be 15 mg/L at a constant temperature.

Graham's Law of Diffusion

The rate of diffusion of a gas is inversely proportional to the square root of its molecular weight. Therefore, the lower the molecular weight of a gas, the faster it will diffuse.

Mathematically,

$$\frac{V_1}{V_2} = \sqrt{\frac{m_2}{m_1}}$$

where:

V_1 = rate of diffusion of a gas, with molecular weight m_1
V_2 = rate of diffusion of a gas, with molecular weight m_2

Assume hydrogen diffuses at a rate of 1,000 m/sec. What will be the rate of diffusion of oxygen? Substituting these values in the formula, the rate of diffusion of oxygen would be 250 m/sec if V_1 and m_1 are of hydrogen and V_2 and m_2 of oxygen.

$$\frac{1{,}000}{V_2} = \sqrt{\frac{32}{2}} \text{ or}$$

$$\frac{1{,}000}{V_2} = 4 \text{ or}$$

$$V_2 = \frac{1{,}000}{4} \text{ m/sec} = 250 \text{ m/sec}$$

Dalton's Law of Partial Pressure

The total pressure of a mixture of gases is the sum of their partial pressures. Gases are generally collected by displacing water in a tube closed at the upper end. This tube is known as an *eudiometer*. Pressure exerted by a gas is a pressure due to gas and water vapor at that temperature. To determine the pressure of the dry gas, the vapor pressure of water is subtracted from the total pressure. Suppose a gas collected over water exerts 750 mm of mercury pressure at 20°C; the pressure due to gas alone or dry gas would be 750 minus 17.5 mm Hg (the water vapor pressure at 20°C) or 732.5 mm of Hg.

Boyle's Law

The volume of a dry gas is inversely proportional to the pressure at a constant temperature. It means the higher the pressure, the lower the volume.

Mathematically,

$$\frac{V_1}{V} = \frac{P}{P_1} \text{ or}$$

$$V_1 = \frac{P}{P_1} \times V$$

where:

V_1 = volume corresponding to pressure P_1
V = volume corresponding to pressure P

Suppose 200 mL of a dry gas exerts 740 mm Hg pressure. What will be the volume if pressure changes to 750 mm Hg?

Substituting these values in the equation:

$$V_1 = \frac{740}{750} \times 200 \text{ mL} = 197.33 \text{ mL}$$

Charles's Law

The volume of a dry gas is directly proportional to the absolute temperature at a constant pressure.

Mathematically,

$$\frac{V_1}{V} = \frac{T_1}{T} \text{ or}$$

$$V_1 = \frac{T_1}{T} \times V$$

where:

V_1 = new volume
T_1 = new temperature
V = original volume
T = original temperature

Example: If a gas measures 500 mL at 20°C, what will be its volume at 0°C?

$$V_1 = \frac{273 + 0}{273 + 20} \times 500 = 466 \text{ mL}$$

Boyle's and Charles's laws combined can be expressed mathematically as

$$\frac{V_1}{V} = \frac{P}{P_1} \times \frac{T_1}{T}$$

Now we can determine any one of the six values by knowing the other five.

Gay-Lussac's Law of Combining Volumes of Gases

Under similar conditions of temperature and pressure, the volumes of reacting gases and of their gaseous products are expressed in a ratio of small whole numbers. This ratio is the same as that of the molecules in the balanced equations.

$$2H_2 + O_2 \rightarrow 2H_2O$$

2 vol : 1 vol : 2 vol

$$HCl + NH_3 \rightarrow NH_4Cl$$

1 vol : 1 vol : 1 vol

Avogadro's Hypothesis

Equal volumes of gases under similar conditions of temperature and pressure contain the same number of molecules. Note that 22.4 L of most of the gases contain an Avogadro number of their molecules (1 mol) at the standard temperature (0°C) and standard pressure (760 mm Hg). Therefore, 22.4 L is known as molar or gram molecular volume. Density of a gas is the mass in grams per litre at standard temperature and pressure (STP); therefore, it is the mole of a gas divided by 22.4 L.

$$\text{Density of } O_2 = \frac{32 \text{ g}}{22.4 \text{ L}} = 1.43 \text{ g/L}$$

Using this principle, the molecular weights of gases may be determined if the mass of some volume of a gas and conditions are specified.

Example: A sample of a certain gas measured at 20°C and 757.5 mm Hg pressure weighs 2 g and its volume is 300 mL. What is its molecular weight? Water vapor pressure at 20°C is 17.5 mm Hg.

Using equation

$$\frac{V_1}{V} = \frac{P}{P_1} \times \frac{T_1}{T}$$

where:

$V = 300$ mL

$P = 757.5 - 17.5$ mm Hg

$P_1 = 760$ mm Hg

$T = 273 + 20$°A

$T_1 = 273$°A

$$\text{Volume at STP} = \frac{740}{760} \times \frac{273}{293} \times 300 = 272 \text{ mL}$$

$$\text{Gram molecular mass} = \frac{2.0 \text{ g}}{272} \times 22.4 \times 1{,}000 = 164 \text{ g}$$

Molecular weight = 164

Questions

1. Give properties of gases.
2. Which one of the following gases will diffuse the fastest?
   a. Chlorine (Cl_2)
   b. Carbon dioxide (CO_2)
   c. Hydrogen sulfide (H_2S)
3. A sample of a gas measures 500 mL at standard conditions. What is its volume at 20°C and 750 mm Hg pressure?

4. Calculate the density of methane (CH_4).
5. Calculate the molecular weight of a gas weighing 3 g and measuring 700 mL at 10°C and 700 mm mercury.

12 Organic Chemistry

This branch of chemistry deals with the compounds of carbon, both natural and synthetic, and their derivatives. Originally, it was limited to substances found in and produced by plants and animals. In 1828, Friedrick Wohler obtained urea, an organic compound, by heating potassium cyanate (KCNO) with ammonium sulfate ($(NH_4)_2SO_4$).

$$2KCNO + (NH_4)_2SO_4 \rightarrow K_2SO_4 + 2\underbrace{H_2N{-}\overset{\overset{\displaystyle O}{\|}}{C}{-}NH_2}_{\text{urea}}$$

Since then, synthesis of various kinds of organic compounds started. Organic chemistry is important in water and wastewater treatment because organics are decomposed in wastewater treatment and removed in water treatment.

As compared to inorganic substances, organics have the following characteristics:

1. They are generally combustible.
2. They have a lower melting point and boiling point.
3. They are less soluble in water.
4. They have isomerism, meaning the existence of a number of compounds with the same molecular formula but different structural formulae.
5. They are mostly covalent compounds.
6. They have high molecular weight, often over 1,000.
7. Most serve as a source of food for bacteria.

Carbon Atoms and Bonding

A carbon atom does not give or take electrons in the compound formation. It shares electrons and forms covalent bonds. It has the tendency to share electrons with other carbon atoms to form chains. That is why organic compounds are covalent with high molecular weight. Each carbon atom has four valence electrons and shares four from other atoms; thus, there are four covalent bonds with each carbon atom in the organic compounds. The angle between every two bonds is 109.28°, and all four bonds are equally strong.

In a chain of carbon atoms, there can be single, double, or triple covalent bonds.

```
    H H
    | |
  H–C–C–H, C2H6 = ethane has single bonds
    | |
    H H
```

```
   H H
   | |
 H–C=C–H, C2H4 = ethene has a double bond
```

H–C≡C–H, C_2H_2 = ethyne or acetylene has a triple bond

Due to isomerism, in organic chemistry, structural formulae rather than molecular formulae are used. As most of the organic compounds are formed of carbon, nitrogen, oxygen, and hydrogen, these elements have four, three, two, and one covalent bonds, respectively.

```
  |
 –C–        –N–        –O–        –H
  |          |
```

For example, in urea:

```
H    O  H
 \   //  /
  N–C–N
 /       \
H         H
```

Classification of Organic Compounds

For our purpose, organic compounds can be divided into four main classes: hydrocarbons, carbohydrates, lipids, and proteins. The last three groups are known as natural organic compounds.

Hydrocarbons

Hydrocarbons are compounds formed of hydrogen and carbon only. They are straight chain, branched chain, or ringed in their molecular structure. Open-chain hydrocarbons are classified into different series, as shown in Table 12-1.

Alkyl Group	Alkane (Paraffin) Series	Alkene (Olefin) Series	Alkyne Series	Alkadiene
Methyl, CH_3^-	Methane, CH_4			
Ethyl, $C_2H_5^-$	Ethane, C_2H_6	Ethene, C_2H_4	Ethyne, C_2H_2	
Propyl, $C_3H_7^-$	Propane, C_3H_8	Propene, C_3H_6	Propyne, C_3H_4	
Butyl, $C_4H_9^-$	Butane, C_4H_{10}	Butene, C_4H_8	Butyne, C_4H_6	Butadiene, C_4H_6
Pentyl, $C_5H_{11}^-$	Pentane, C_5H_{12}	Pentene, C_5H_{10}	Pentyne, C_5H_8	Pentadiene, C_5H_8
Hexyl, $C_6H_{13}^-$	Hexane, C_6H_{14}	Hexene, C_6H_{12}	Hexyne, C_6H_{10}	Hexadiene, C_6H_{10}

Table 12-1 Hydrocarbon Series

Alkane Series

The alkane series are straight- or branched-chain hydrocarbons with single covalent bonds in between carbon atoms. The general formula for them is C_nH_{2n+2} where n = number of C atoms in the chain.

```
                  H H H H
                  | | | |
n. Butane,      H–C–C–C–C–H, C4H10
                  | | | |
                  H H H H
```

Because the most common bonding of carbon atoms is with hydrogen atoms, sometimes hydrogen is not written in the formula; e.g., methane,

```
  H
  |
H–C–H
  |
  H
```

may be written as

```
 |
–C–
 |
```

An unbranched chain is indicated by "n." in front of the name of the hydrocarbon, e.g., n. butane.

Alkene Series

The alkene series are straight- or branched-chain hydrocarbons in which there is a double bond in the molecule. For example:

```
                  H H
                  | |
Ethene (C2H4) is H–C=C–H
```

Hexene (C_6H_{12}) is

```
    H H H H H H
    | | | | | |
  H-C-C-C=C-C-C-H
    | |     | |
    H H     H H
```

A double bond will reduce the number of hydrogen atoms by two; therefore, the general formula for the alkene series is C_nH_{2n}.

Alkyne Series

The alkyne series are straight- or branched-chain hydrocarbons with a triple bond in between carbon atoms in the chain. Because a triple covalent bond reduces the number of hydrogen atoms by 4, the general formula is C_nH_{2n-2}.

Ethyne or acetylene, H–C≡C–H, C_2H_2

Alkadiene Series

The alkadiene series differ from the other series by having two double bonds in the chain. For example, butadiene is

```
H   H H   H
 \  | |  /
  C=C-C=C, C4H6
 /       \
H         H
```

Its general formula, like alkyne, is C_nH_{2n-2}.

Benzene Series or Aromatic Series

These hydrocarbons have a benzene ring(s). A benzene ring is a hexagonal ring, with each corner represented by a carbon atom with alternating double bonds.

H
C
Benzene (C_6H_6), H—C C—H or simply
H—C C—H
C
H

Benzene is prepared commercially by the distillation of coal tar. Because benzene has a pleasant aromatic odor, its compounds are known as aromatic hydrocarbons (e.g., naphthalene and anthracene).

H H
C C
H—C C C—H
or simply
H—C C C—H
C C
H H

Naphthalene, $C_{10}H_8$

H H H
C C C
H—C C C C—H
H—C C C C—H
C C C
H H H

Anthracene, $C_{14}H_{10}$

When other groups replace hydrogen atoms in benzene molecules, the compounds are named as ortho, meta, and para, for bonding with consecutive carbon atoms, alternate carbon atoms, and opposite carbon atoms, respectively. For example, two methyl groups (CH_3^-) with the benzene ring form a xylene that is an ortho, meta, and para type depending on the position of the bonds.

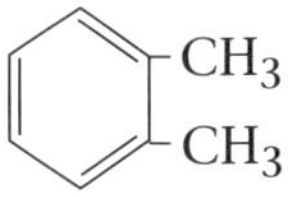

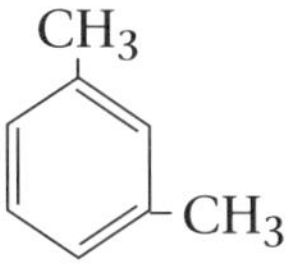

Ortho-xylene Meta-xylene Para-xylene

The molecule of a hydrocarbon with one hydrogen replaced is known as an alkyl group in an alkane series and a phenyl group in a benzene series. They are represented in the general formula by "R."

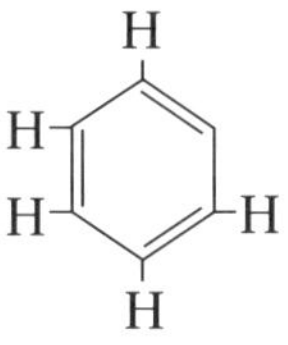

Phenyl ($C_6H_5^-$)

Derivatives of Hydrocarbons

These compounds are produced by the substitution of hydrogen in the hydrocarbon compounds by other groups known as *functional groups* (shown in Table 12-2).

Alcohols

They are hydrocarbons with one or more hydrogens replaced by a hydroxyl (–OH) group. Their names end with *-nol*.

Methanol, H–C(H)(H)–OH , CH_3OH

Derivative	General Formula	Name (Trivial, IUPAC*)
Alcohol	R–OH	Ethyl alcohol, Ethanol
Ether	R–O–R	Diethyl ether, Ethoxyethane
Aldehyde	$R{-}C({=}O){-}H$	Acetyldehyde, Ethanal Formaldehyde, Methanal
Ketone	$R{-}C({=}O){-}R$	Acetone, Propanone
Acid†	$R{-}C({=}O){-}O{-}H$	Formic acid, Methanoic acid Butyric acid, Butanoic acid Valeric acid, Pentanoic
Ester†	$R{-}C({=}O){-}O{-}R$	Ethyl acetate Glyceryl trinitrate
Amines†	$R{-}NH_2$	Propylamine Methyl ethyl amine Methyl ethyl n. propylamine
Amides†	$R{-}C({=}O){-}N(H){-}R$	Acetamide
Chloroorganics†	R–Cl(s)	Trichloromethane Chlorobenzene

* International Union of Pure and Applied Chemistry.
† Have trivial names, generally.

Table 12-2 Derivatives of Hydrocarbons

$$\text{Ethanol,}\qquad H{-}\underset{H}{\overset{H}{C}}{-}\underset{H}{\overset{H}{C}}{-}OH\ ,\ C_2H_5OH$$

Ethers

They are organic oxides. They are derivatives of water in which both hydrogen atoms are replaced by alkyl groups.

$$\begin{array}{ccccccccccc} & & H & & H & & & H & & H & \\ & & | & & | & & & | & & | & \\ H & - & C & - & C & - & O - & C & - & C & - H \\ & & | & & | & & & | & & | & \\ & & H & & H & & & H & & H & \end{array} , \ C_2H_5OC_2H_5$$

Diethyl ether or ethoxyethane

Aldehydes

They have a hydrogen of the hydrocarbon replaced by the formyl groups,

$$\begin{array}{c} O \\ // \\ -C-H \end{array}$$

Methanal or formaldehyde $\begin{array}{c} O \\ // \\ H-C-H \end{array}$, CH_2O

Their technical names end with *-al*, e.g., methanal and ethanal.

Ketones

These compounds have hydrocarbon hydrogen, replaced by a carbonyl group

$$\begin{array}{c} O \\ // \\ -C- \end{array}$$

Their names end with *-one*. For example, propanone or acetone is

$$\begin{array}{c} O \\ // \\ H_3C-C-CH_3 \end{array}$$

Acids

All organic acids have a carboxylic group (–COOH) substituting hydrogen atom(s). Their names end with *-noic*. For example:

Methanoic (formic) acid, $H{-}C({=}O){-}OH$

Butanoic acid, $H_7C_3{-}C({=}O){-}OH$

Esters

They are produced when acids react with alcohols. A water molecule is produced when an acid group reacts with an alcohol group. Bonding between an acid and alcohol molecule is known as *esterification*. Ethyl acetate is the product of ethyl alcohol and acetic acid.

$$C_2H_5 - OH + CH_3COOH \longrightarrow CH_3COOC_2H_5 + H_2O$$

$$CH_3COO\ C_2H_5 = H{-}\underset{H}{\overset{H}{C}}{-}\overset{O}{\overset{\|}{C}}{-}O{-}\underset{H}{\overset{H}{C}}{-}\underset{H}{\overset{H}{C}}{-}H$$

Esterification

Amines

They are derivatives of ammonia in which one, two, or all three hydrogen atoms of the ammonia molecule are replaced by alkyl groups. For example, propylamine is

$$C_2H_5{-}N{<}^{H}_{H}\ ,\ C_2H_5NH_2$$

Amides

They are formed when an amine is bonded with a carboxylic group (–COOH) of an acid. For example, in proteins where –COOH of one amino acid is bonded with the amine (NH_2–) group of another. The bond formed is known as an amide bond:

```
          H       O H   CH3
          |      //  |   |
   H2N – C – C – N – C – COOH
          |      ↑       |
          H      |       H
             Amide bond
        Glycine     Alanine
```

Chloroorganics

A large number of chloroorganic compounds are formed when chlorine reacts with hydrocarbons and replaces hydrogen. They have health effects and are contaminants of drinking water (see Figure 12-1). Some important chloroorganic classes are volatile organics and chlorine disinfection by-products.

Carbohydrates

Carbohydrates contain the elements carbon, hydrogen, and oxygen. Hydrogen and oxygen atoms in all carbohydrates are in a 2:1 ratio as in water; therefore, the name is carbohydrates. Sugars, starches, celluloses, and substances closely related to them are all naturally occurring carbohydrates. Most common carbohydrates have proved to be condensation polymers (formed of several molecules of a simple compound) of simpler molecular sugar units, known as monosaccharides, or simple sugars. Glucose, galactose, and fructose are simple sugars. They are isomers with a molecular formula of $C_6H_{12}O_6$. Formulas are simplified by omitting C and H, except in glucose, for clarification.

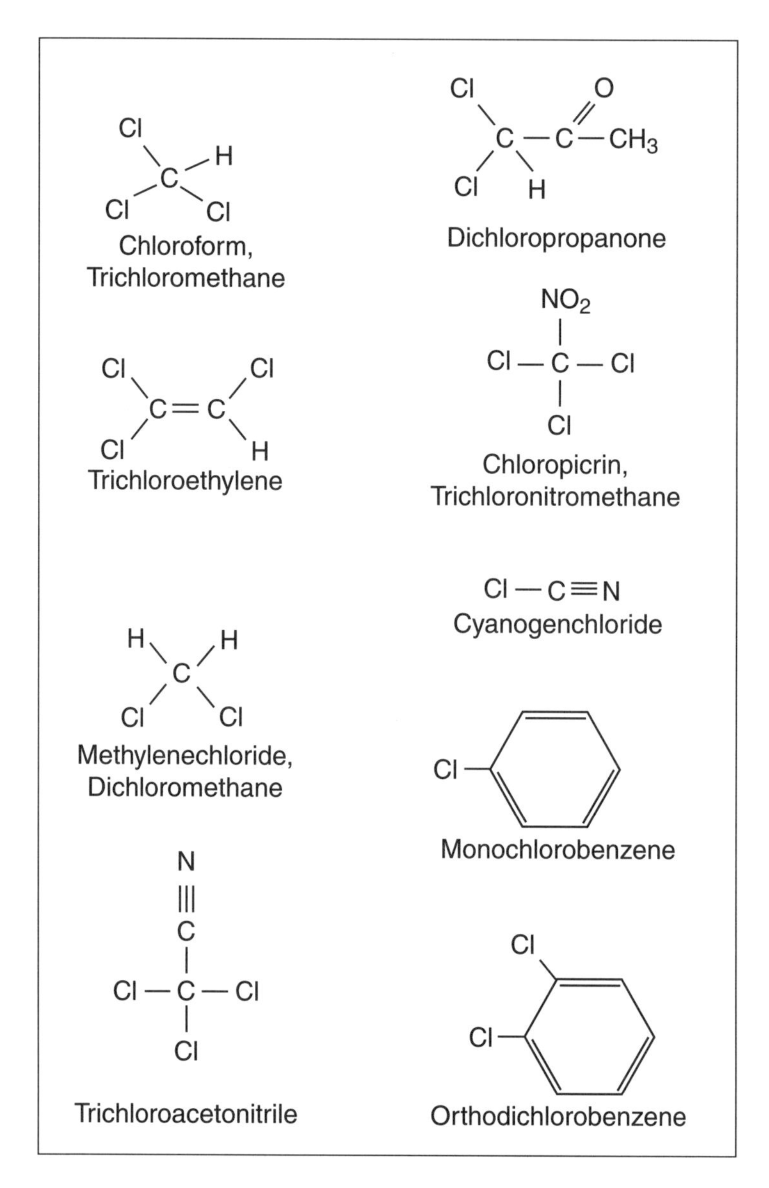

Figure 12-1 Some Chloroorganic Contaminants of Drinking Water

Glucose

Galactose

Fructose

Disaccharides

Two monosaccharides combine to form a disaccharide and a water molecule. Examples of disaccharides are sucrose, maltose, and lactose. Glucose and fructose form sucrose, glucose and glucose form maltose, and glucose and galactose form lactose.

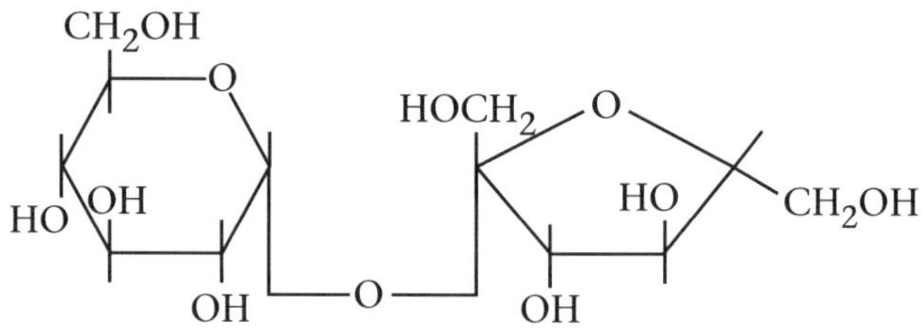

Sucrose

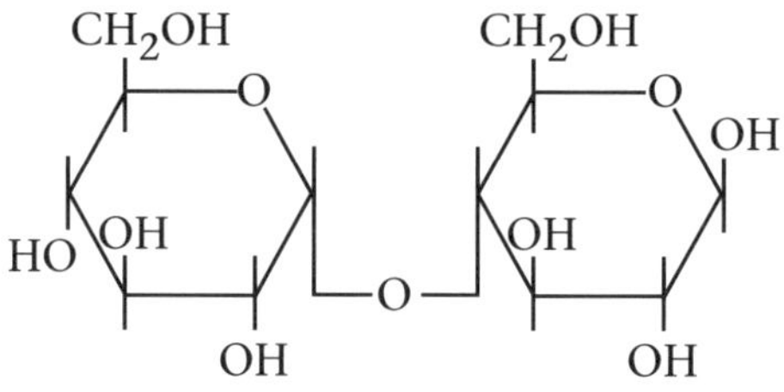

Maltose

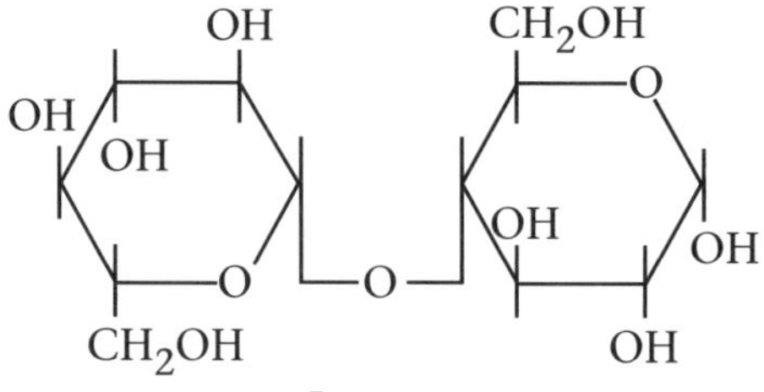

Lactose

Polysaccharides

Polysaccharides are formed when several monosaccharides combine together. Three common polysaccharides are starch, cellulose, and glycogen. They are polymers of glucose.

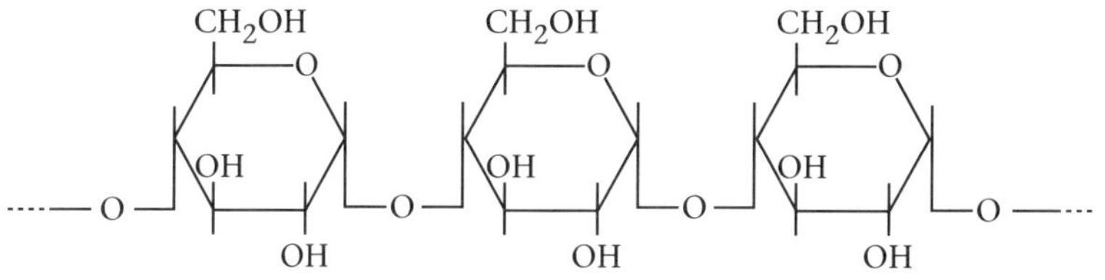

Starch: A Polymer of "α" Glucose

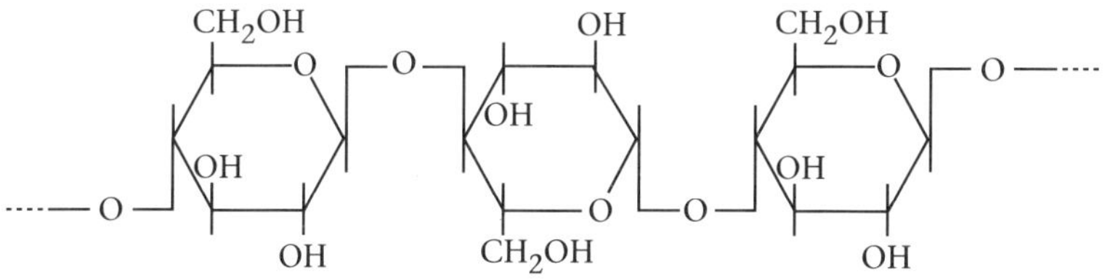

Cellulose: A Polymer of "β" Glucose

Disaccharides yield two simple sugars by taking one molecule of water. Polysaccharides react with water to produce many monosaccharides. This decomposition with water is known as hydrolysis.

Lipids

Lipids are mainly fats and oils that contain the same elements (C, H, O) as carbohydrates but have comparatively less oxygen. Upon hydrolysis, fats yield fatty acids and glycerol. They are esters of glycerol and long-chain fatty acids, usually with an even number of C atoms, from 12–20. Lipids are found in animal and plant tissue. They are insoluble in water and soluble in solvents, such as ether, chloroform, or benzene.

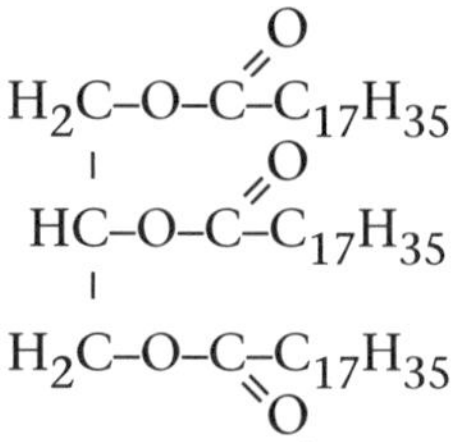

$$\begin{array}{ll}
H_2C{-}O{-}C({=}O){-}C_{17}H_{35} & H_2C{-}O{-}C({=}O){-}C_{17}H_{33} \\
\quad | & \quad | \\
HC{-}O{-}C({=}O){-}C_{17}H_{35} & HC{-}O{-}C({=}O){-}C_{17}H_{33} \\
\quad | & \quad | \\
H_2C{-}O{-}C({=}O){-}C_{17}H_{35} & H_2C{-}O{-}C({=}O){-}C_{17}H_{33}
\end{array}$$

Stearin, a fat
Glycerol + Stearic Acid
melting point = 71°C

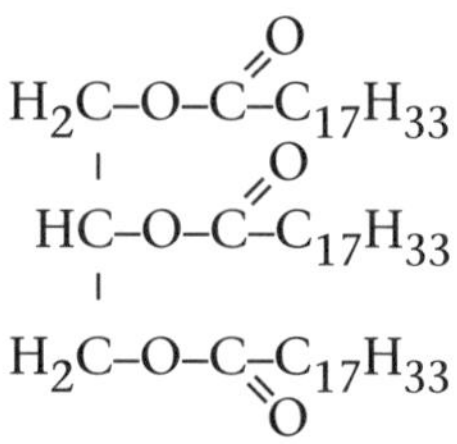

Olein, an oil
Glycerol + Oleic Acid
melting point = –4°C

The molecular structure in fats and oils is the same. Oils are liquid at room temperature due to the presence of some double bonds (unsaturated) in the carbon chain of their fatty acids that lowers the melting point. Fats have single bonds (saturated) and are solid at room temperature.

Saponification

Saponification is the hydrolysis of a lipid by using a solution of a strong hydroxide to form soaps.

$$\begin{array}{l}
H_2C{-}COO{-}R \\
\quad | \\
HC{-}COO{-}R_1 + 3NaOH \rightarrow \underset{\text{soap}}{RCOONa} + R_1COONa + R_2COONa + C_3H_5(OH)_3 \\
\quad | \\
H_2C{-}COO{-}R_2
\end{array}$$

Soaps are generally made by hydrolyzing fats and oils with superheated water at about 250°C under a pressure of about 50 atmospheres. The long-chain carboxylic acids formed are neutralized with sodium hydroxide to yield a mixture of sodium salts, which constitutes soap.

Hydrogenation

Hydrogenation is the conversion of an oil into a fat. This is achieved by adding hydrogen to the unsaturated fatty acid and making it saturated. The hydrogenation is controlled so that the resulting product is not completely saturated.

$$\cdots -\overset{|}{\underset{|}{C}}-\overset{|}{\underset{|}{C}}=\overset{|}{\underset{|}{C}}-\overset{|}{\underset{|}{C}}- \cdots \qquad\qquad \cdots -\overset{|}{\underset{|}{C}}-\overset{|}{\underset{|}{C}}-\overset{|}{\underset{|}{C}}-\overset{|}{\underset{|}{C}}- \cdots$$

$$\underset{\text{Oleic acid}}{C_{17}H_{33}\,COOH + H_2} \rightarrow \underset{\text{Stearic acid}}{C_{17}H_{35}\,COOH}$$

Proteins

Proteins contain carbon, hydrogen, oxygen, and nitrogen. Some proteins also contain sulfur (S) and phosphorus (P) as additional elements. They are known as nitrogenous organics. They are formed of chains of amino acids linked by amide bonds. Hydrolysis of proteins yields amino acids. An amino acid contains both $-NH_2$ and $-COOH$ groups.

The general formula of an amino acid and a protein is

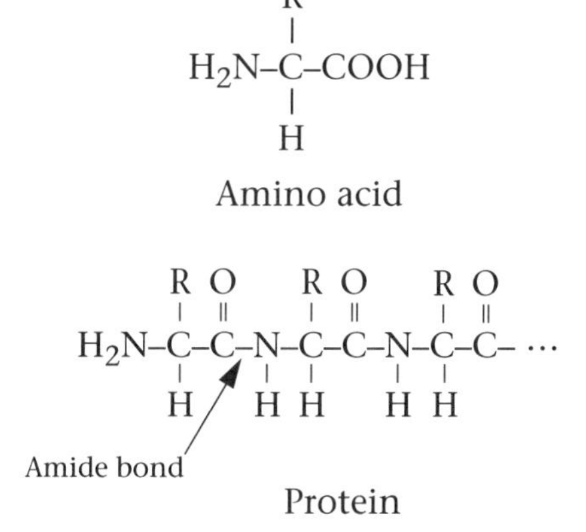

Amino acid

Protein

Proteins form two-thirds of the dry weight of all living cells. While fats and carbohydrates are used mainly as sources of energy, proteins are involved in a variety of living activities of the cell. Some examples of proteins are the white of an egg, enzymes, hormones, and antibodies. They are the main component of nails, hair, and muscles. Proteins have a key position in the architecture and functioning of living matter. *Protein* means *preeminent* or *first* in Greek.

Classification of Proteins

Proteins can be classified into three groups:

Simple proteins: They yield only amino acids on hydrolysis, e.g., egg albumin.

Conjugated proteins: They are combined with a characteristic nonamino acid substance, e.g., glycoproteins, lipoproteins, nucleoproteins, and phosphoproteins.

Derived proteins: They are the products derived from either the simple or conjugated proteins. Proteins, when digested, form peptides, amino acids, and ammonia.

Polypeptides → Dipeptides → Amino acids → Ammonia

The amount of heat produced by the oxidation of fats, carbohydrates, and proteins is approximately 9.3, 4.1, 4.1Kcal/g, respectively. Lipids and carbohydrates are oxidized completely to H_2O and CO_2, while proteins undergo incomplete oxidation and some end products still contain C and H and, thus, some energy. End products of proteins are ammonia and hydrogen sulfide, which are finally oxidized to nitrates and sulfates.

Questions

1. Give the differences between organic and inorganic compounds.
2. Define:
   a. Isomerism
   b. Ether
   c. Esterification
   d. Saponification

3. Give the general formulae of the following:
   a. Alcohol
   b. Carboxylic acid
   c. Aldehyde
   d. Ketone
   e. Amino acid
4. Differentiate:
   a. Carbohydrate from a hydrocarbon
   b. Fat from oil
   c. Monosaccharides from disaccharides
   d. Simple protein from a conjugated protein

Hardness

Hardness is caused by the presence of dissolved calcium and magnesium compounds. In natural waters, it varies considerably depending on the source. Areas with limestone ($CaCO_3$) and dolomite ($CaCO_3$ and $MgCO_3$) deposits have more hardness than others. Hardness results in an increased consumption of soap. The harder the water, the higher the consumption of soap, and vice versa. Commonly, hardness is caused by the bicarbonates (HCO_3^-), sulfates ($SO_4^=$), nitrates (NO_3^-), and chlorides (Cl^-) of calcium and magnesium.

Rainwater usually contains carbon dioxide, CO_2, in solution as carbonic acid, H_2CO_3. As this water percolates through deposits of limestone and dolomite, some of these carbonates are converted into soluble bicarbonates.

$$CaCO_3 + H_2CO_3 \rightarrow Ca(HCO_3)_2$$

Rainwater also dissolves sulfates and chlorides of calcium and magnesium. The presence of other metals such as iron (Fe) and manganese (Mn) may also cause hardness. Waters are classified into various types depending on the degree of hardness.

Hardness	Water
15–50 mg/L as $CaCO_3$*	Soft
50–100 mg/L as $CaCO_3$	Medium hard
100–200 mg/L as $CaCO_3$	Very hard

*Units mg/L (ppm, parts per million) are expressed as $CaCO_3$, meaning equivalent to $CaCO_3$.

Lather Formation and Hardness

Common washing soap is a water-soluble salt of sodium and a fatty acid, e.g., sodium stearate ($NaC_{18}H_{35}O_2$).

When soap is added to hard water, it reacts with the calcium and magnesium compounds to form insoluble compounds, the scum or curd, that deposit on clothes.

$$CaSO_4 + 2NaC_{18}H_{35}O_2 \rightarrow \underset{\text{scum}}{Ca(C_{18}H_{35}O_2)_2\downarrow} + Na_2SO_4$$

This reaction continues until all the calcium and magnesium ions are precipitated. Soft water does not contain these ions and, therefore, lathers easily when soap is added.

Types of Hardness

1. Carbonate or temporary hardness is commonly due to bicarbonates (HCO_3^-) of calcium and magnesium and sometimes due to their carbonates. It is common in areas with limestone and dolomite deposits. It is considered "temporary" because it can be removed by boiling the water.
2. Noncarbonate or permanent hardness is caused by soluble compounds such as sulfates ($SO_4^=$), nitrates (NO_3^-), and chlorides (Cl^-) of calcium and magnesium. These compounds are more stable in their solutions than bicarbonates. They cannot be removed by boiling the water.

Problems Due to Hardness

1. It is a nuisance in laundering, due to the wasting of soap and the collecting of precipitate on fibers.
2. It is a nuisance in bathing.

3. It causes the formation of a scale inside steam boilers. The scale is formed by the carbonate hardness. It may act as a heat insulator, preventing proper transfer of heat.

Softening Methods

Boiling

Boiling removes carbonate hardness. Bicarbonates are converted into insoluble carbonates.

$$Ca(HCO_3)_2 \rightarrow CaCO_3\downarrow + H_2O + CO_2\uparrow$$

Precipitation

Precipitation is the addition of basic solutions such as $Ca(OH)_2$ or NH_4OH to form $CO_3^{=}$ and H_2O from bicarbonates. Magnesium carbonate ($MgCO_3$) is further reacted with $Ca(OH)_2$ to precipitate magnesium hydroxide ($Mg(OH)_2$).

$$Ca(HCO_3)_2 + Ca(OH)_2 \rightarrow 2CaCO_3\downarrow + 2H_2O$$

$$Mg(HCO_3)_2 + Ca(OH)_2 \rightarrow MgCO_3 + CaCO_3\downarrow + 2H_2O$$

$$MgCO_3 + Ca(OH)_2 \rightarrow Mg(OH)_2\downarrow + CaCO_3\downarrow$$

The removal of magnesium carbonate hardness requires double the amount of lime and pH above 10.6; whereas, calcium is removed above pH 9.4.

Lime ($Ca(OH)_2$) and sodium carbonate (Na_2CO_3) or soda ash are added to remove noncarbonate hardness. Calcium and magnesium precipitate out as insoluble carbonate and hydroxide, respectively.

$$CaSO_4 + Na_2CO_3 \rightarrow CaCO_3\downarrow + Na_2SO_4$$
$$MgSO_4 + Ca(OH)_2 \rightarrow Mg(OH)_2\downarrow + CaSO_4$$

This method is known as lime–soda ash softening and is commonly used in water treatment. Because soda ash is a relatively expensive chemical and waters with less than 50 mg/L of residual hardness can seldom be produced, it is normal to leave about two-thirds of the residual hardness in the permanent form using the lime–soda ash process. Therefore, the soda ash requirement may be calculated as follows:

$$\text{Noncarbonate hardness} - \tfrac{2}{3}\ \text{expected residual hardness} \times \frac{106}{100} = Na_2CO_3 \text{ dose required as mg/L.}$$

Na_2CO_3 dose as mg/L is ⅓ permanent hardness × 1.06.

It is impossible to produce waters with less than 25 mg/L hardness by the lime–soda ash method because $Mg(OH)_2$ has a solubility of 9 mg/L and $CaCO_3$, 17 mg/L. Proper detention time is not practical; therefore, in waters softened by this method, there is normally 50–80 ppm (mg/L) hardness.

Ion Exchange

Certain natural minerals, known as zeolites (boiling stone), have a porous, three-dimensional network of alumino–silicate groups that act as large, fixed ions carrying a negative charge. They occur in cavities and cracks of rocks, such as granite and basalt. Metallic ions such as Na^+ are attached to these ions to form large molecules. If hard water is allowed to stand in sodium zeolite, Ca^{++} and Mg^{++} ions replace Na^+ ions.

$$CaSO_4 + Na_2Z \rightarrow CaZ + Na_2SO_4$$

where:

Z = zeolite

An exhausted zeolite is regenerated by immersing it in a strong solution of sodium chloride, NaCl. Presently, synthetic zeolites are replacing natural zeolites. Also, ion exchange resins are used for the same purpose. Resins act similar to zeolites. The reactions are

$$Ca(HCO_3)_2 + H_2R \rightarrow CaR + 2H_2O + 2CO_2\uparrow$$
$$CaSO_4 + H_2R \rightarrow CaR + H_2SO_4$$

where:

R = resin

These resins are regenerated with acids. Natural water treated by a combination of cation and anion exchange resins is called deionized or demineralized water, which is required sometimes for industrial use and in laboratories.

Iron and Manganese in Water

Iron and manganese are commonly present in water with low pH and a lack of molecular oxygen. Many water supplies contain quantities of iron and manganese that cause them to be unsatisfactory for domestic or industrial use without treatment. Both iron and manganese can be present in all groundwaters. Waters containing more than 0.3 ppm of iron and 0.05 ppm of manganese are objectionable.

Problems Caused by Iron and Manganese

1. Iron and manganese in water will stain clothes yellow and enamel black.
2. Their oxides are undesirable for bottling, laundries, paper mills, tanning, and ice manufacturing.
3. They cause taste and color problems. Iron in water spoils the taste and color of vegetables.
4. Iron deposits as oxides in distribution systems and in water meters, causing the latter to give inaccurate readings.

Mostly, these metals are present as bicarbonates due to the action of carbonic acid, H_2CO_3.

$$Fe + 2H_2CO_3 \rightarrow Fe(HCO_3)_2 + H_2\uparrow$$

$$Mn + 2H_2CO_3 \rightarrow Mn(HCO_3)_2 + H_2\uparrow$$

Sometimes their sulfates are also present due to the bacterial action on sulfides forming H_2S, which gets converted into H_2SO_4.

Removal

These metals can be removed by several methods depending upon the nature of the water.

Aeration

Aeration is the mixing of air in water to dissolve oxygen and to remove carbon dioxide. It is done before softening for effective precipitation of these metals.

The removal of carbon dioxide (CO_2) raises the pH because CO_2 in water is carbonic acid (H_2CO_3). For effective iron removal, pH should be raised to 7.5.

Aeration Process

$$Fe(HCO_3)_2 \rightarrow FeO + 2CO_2\uparrow + H_2O$$

$$4FeO + O_2 \rightarrow \underset{\text{Rust}}{2Fe_2O_3\downarrow}$$

$$Mn(HCO_3)_2 \rightarrow MnO + 2CO_2\uparrow + H_2O$$

$$6MnO + O_2 \rightarrow \underset{\text{an unstable compound}}{2Mn_3O_4}$$

$$4Mn_3O_4 + O_2 \rightarrow 6Mn_2O_3\downarrow$$

Manganese oxidizes slower than iron, and higher pH is required for its removal. Manganous oxide will rapidly oxidize to Mn_3O_4, which slowly converts into a precipitate of Mn_2O_3. Both Fe_2O_3 and Mn_2O_3 have $3H_2O$ as water of hydration and, therefore, are also written as $Fe(OH)_3$ and $Mn(OH)_3$.

$$Fe_2O_3 \cdot 3H_2O \text{ or } 2Fe(OH)_3$$

Chemical Precipitation

Iron is commonly present in well water as ferrous iron, which can be removed by increasing pH to 9.4 with lime.

$$4Fe(HCO_3)_2 + 8Ca(OH)_2 + O_2 \rightarrow 2Fe_2O_3\downarrow + 8CaCO_3\downarrow + 12H_2O$$

Manganese will also be removed to some extent by $Ca(OH)_2$. It requires a higher pH.

$$2Mn(HCO_3)_2 + 4Ca(OH)_2 + O_2 \rightarrow 2MnO_2\downarrow + 4CaCO_3\downarrow + 6H_2O$$

Manganese dioxide (MnO_2) is comparatively less soluble.

Chlorination

Chlorine also helps in the removal of iron and manganese, especially if the water contains microbes.

$$3MnO + HOCl \rightarrow Mn_3O_4 + HCl$$

$$2Mn_3O_4 + HOCl \rightarrow 3Mn_2O_3\downarrow + HCl$$

$$2FeO + HOCl \rightarrow Fe_2O_3\downarrow + HCl$$

With Zeolites

Removal of iron and manganese by the cation exchange method is similar to and simultaneous with the removal of calcium and magnesium. In order to avoid the accumulation of ferric hydroxide in the zeolite bed, it is important that all iron in water be in the ferrous (Fe^{+2}) form. Therefore, there should not be any aeration of water, and iron contents should not be very high.

For iron and manganese removal, zeolites are pretreated with manganous sulfate ($MnSO_4$) and potassium permanganate ($KMnO_4$), respectively.

Questions

1. Name the elements that cause hardness in water.
2. Differentiate carbonate hardness from noncarbonate hardness.
3. Give various methods for removing hardness.
4. Give various problems caused by hardness in water.
5. Why is removal of magnesium hardness more expensive than that of calcium?
6. Give acceptable limits of the amounts of Fe and Mn in drinking water.

14 Corrosion and Its Control

Corrosion is the deterioration of a substance due to its reaction with the environment. In water systems, it is the dissolving of metals, such as iron, lead, or copper. It is a natural process that returns processed metals back to their natural state as minerals. Corrosion is essentially an electrochemical reaction formed of three parts: the anode, the cathode, and an electrolyte. The *anode*, the positive electrode, is the site where the metal is dissolving by losing electrons. The *cathode*, the negative electrode, is where an oxidizing agent accepts the electrons. The *electrolyte* is the conducting medium, which is water. The loss of electrons is the oxidation and the increase in electrons is the reduction; therefore, corrosion is a redox reaction. A reaction will not take place unless all three of its parts are present. Reducing agents are mostly metals, such as iron, lead, and copper, which are anodic. Oxidizing agents are oxygen, hydrogen ions, and other metals, which become cathodic. Here is an example of the corrosion of iron in water.

Anode (positive electrode)

$$Fe \rightarrow Fe^{+2} + 2e^-$$

Cathode (negative electrode)
In the presence of O_2

$$\tfrac{1}{2} O_2 + H_2O + 2e^- \rightarrow 2(OH^-)$$

Overall reaction is

$$Fe^{+2} + 2(OH^-) \rightarrow Fe(OH)_2$$

All of these reactions can be summarized as

$$2Fe + O_2 + 2H_2O \rightarrow 2Fe(OH)_2$$

Under acidic conditions

$$2H^+ + 2e^- \rightarrow H_2\uparrow$$

Ferrous hydroxide ($Fe(OH)_2$) is soluble. But in the presence of oxygen, it is further oxidized into ferric hydroxide ($Fe(OH)_3$), which is rusty in color and insoluble.

$$4Fe(OH)_2 + O_2 + 2H_2O \rightarrow \underset{\text{Rusty slurry}}{4Fe(OH)_3\downarrow}$$

At the dead ends, the oxygen is depleted by these reactions forming the rusty slurry. Some of this rusty "slurry" is swept from the anode site by higher water velocities, causing rusty water complaints. As corrosion proceeds, a pit may form at the anodic site with blackish gray $Fe(OH)_2$ inside and an insoluble, rusty ($Fe(OH)_3$) layer on the outside. This outside growth is known as a *tubercle*. High velocity or diffusion of water into the tubercle can rupture it and release $Fe(OH)_2$, which is readily oxidized to rust ($Fe(OH)_3$).

Types of Corrosion

There are many ways to classify corrosion, but mostly it is based on its forms and causes.

1. *Physical corrosion:* It is erosion due to high-velocity (i.e., over 5 ft/sec), particulate matter, or when dispersed gas bubbles erode the surface of a pipe.
2. *Stray current corrosion:* It occurs as a localized attack caused by the potential differential from an outside source, such as grounding appliances, through pipes. Corrosion occurs where the current leaves the pipe.
3. *Uniform corrosion:* It takes place at an equal rate all over the surface of the metal due to the uniform

presence of many electrochemical cells throughout the surface. Mostly this is associated with low pH of the water. Any one site on the metal surface may be anodic one instant and cathodic the next. The loss in weight is directly proportional to the time of exposure.

4. *Localized or pitting corrosion:* It occurs where there is an exposed area of an otherwise coated metal surface due to stress. This localized area becomes anodic and is surrounded by a large cathodic area. The pit remains anodic and becomes larger and larger as corrosion progresses. Often, the pit is covered with a tubercle.
5. *Concentration cell corrosion:* It is a function of the difference in concentration of various dissolved substances, such as metal and hydrogen ions or oxygen molecules, on the adjacent portions of the metal surface. The corrosion process will tend to equalize the metal ion concentration by dissolving the areas with less concentration, which become anodic. Similarly, the part with less oxygen acts anodic, the one with more oxygen acts cathodic.
6. *Galvanic corrosion:* This occurs when two dissimilar metals are connected together in the water lines. One metal becomes anodic and the other cathodic. A metal on top (more active) in an electromotive series becomes an anode and the one below, the cathode. For example lead-solder and copper pipe will have lead as the anode and copper as the cathode, causing the corrosion of lead.
7. *Bacterial corrosion:* At dead ends of pipes where anaerobic conditions exist (lack of oxygen), sulfates are converted into hydrogen sulfide, H_2S, by anaerobic bacteria. Because H_2S is an acid, it increases H^+ concentration and dissolves iron, which causes very unpleasant tastes and odors.

Various Factors Affecting Corrosion

Dissolved Solids

Different types of ions and their concentrations have different effects on corrosion. Carbonates, polyphosphates, and silicates normally reduce corrosion by forming a protective film on the metal surface; whereas, chlorides may increase corrosion by creating acidic conditions that interfere with the protective film.

Dissolved Gases

Carbon dioxide and oxygen are the most common dissolved gases. Carbon dioxide forms carbonic acid (H_2CO_3) in water. It is an amphiprotic compound with an important buffering effect in water. Oxygen, as previously discussed, acts cathodic and as a depolarizer as it combines with hydrogen to form water. A higher concentration of oxygen causes more corrosion, and vice versa. Other gases of interest are chlorine, ammonia, and hydrogen sulfide. Chlorine as hypochlorous acid causes acidic conditions and thus more corrosion. Hydrogen sulfide is also an acid and thus causes corrosion. Ammonia forms chloramines with chlorine, which are known to inhibit corrosion.

Temperature

As a rule, the higher the temperature, the higher the rate of corrosion. There is less corrosion during winter than summer. Soft waters (with alkalinity less than 50 mg/L) are very corrosive during summer.

Corrosion Control

As previously discussed, the corrosion process has three basic parts and all are important for a reaction to take place. By preventing any one of these three parts, corrosion can be prevented. The slowest part determines the rate of corrosion. The cathodic reaction is generally the slowest due to uncertain availability of oxygen or hydrogen ions. If the cathode surface is covered with OH^- ions and H_2 molecules, it creates a barrier for further reaction, and corrosion is controlled. This physical barrier is known as *polarization.* The disruption or removal of this barrier is called *depolarization,* which causes the resumption of corrosion. Depolarization can be caused by increasing acidity, which neutralizes OH^- ions; by increasing oxygen concentration, which combines with hydrogen to form water; or by increasing the water velocity, which sweeps away OH^- ions and hydrogen gas. Nitrates and chlorine are also depolarizers.

In fact, almost any metal in contact with water will corrode. In many cases, the electron acceptor is the H^+ ion produced by water itself due to self-ionization.

$$H_2O \rightleftarrows H^+ + OH^-$$

Therefore, for full protection, the metal should not be allowed to come in direct contact with water. This can be achieved either by providing a mechanically applied coating, such as a cement lining, with the tar coating, paint, plastic, rubber, etc., or by providing a water that coats the surface of the pipe. Coatings formed by contact with flowing water include calcium carbonate ($CaCO_3$), a phosphate, or a silicate coating.

$CaCO_3$ coating size and rate depend mainly upon the pH of the water and its calcium carbonate contents. At a high pH and a high calcium carbonate concentration, there can be too much $CaCO_3$ deposition (scale formation) in the filter media and the lines. At a low pH and a low calcium carbonate concentration, there can be little or no deposition. For this reason, water is treated to be slightly depositing. There are two general methods of stabilizing hard waters softened by the lime and soda ash method. First is the use of a small amount of polyphosphates, such as sodium hexametaphosphate $6(NaPO_3)$, to prevent too much precipitation of $CaCO_3$ by sequestering. The second method is recarbonation, which is the passing of carbon dioxide (CO_2) through the water to lower its pH and thus convert insoluble $CaCO_3$ into soluble bicarbonate (HCO_3^-).

Various indices have been developed to determine the stability of water based on various factors affecting the $CaCO_3$ deposition characteristics of water.

Various Indices

1. *Marble test:* This test is based on the alkalinity of the water. The water to be tested is saturated with $CaCO_3$ by adding $CaCO_3$ powder to it, shaking it, and keeping it overnight. Alkalinity of the sample is determined before and after saturation. If the initial alkalinity is equal to the final alkalinity, the water is stable. If the initial alkalinity is higher, then water is depositing; if it is less, then it is corrosive. This is a simple, easy, and good way for an operator to control corrosion.
2. *Baylis curve:* This shows the solubility of $CaCO_3$ with regard to alkalinity and pH (see Figure 14-1). If the point of intersection of pH and alkalinity of the water is above the equilibrium curve, the water is depositing; if it is below, the water is corrosive; and if it is equal, the water is stable.
3. *Langelier saturation index (LSI):* This is a commonly used index in the water utility industry (see Figure 14-2). It determines the $CaCO_3$ deposition property of the water by calculating saturation pH (pH_s). pH_s is calculated from total dissolved solids, temperature, alkalinity, and calcium contents of the water. If the pH of the water is equal to the pH_s, the

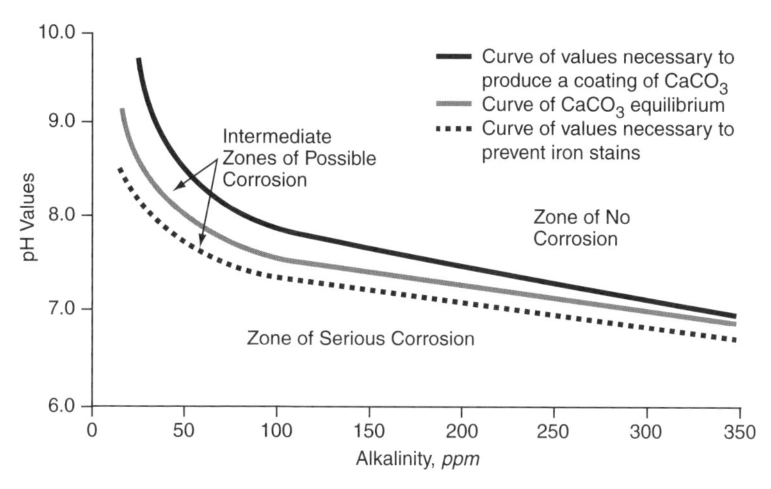

Figure 14-1 Stability Curve Showing Relationship Between pH Values and Alkalinity

water is stable; if it is higher, the water is depositing; and if it is less, the water is corrosive.

$$\text{LSI} = \text{pH} - \text{pH}_s$$

4. *Ryzner index (RI) or stability index:* This index is two times pH_s minus pH of the water.

$$\text{RI} = 2\text{pH}_s - \text{pH}$$

An RI value less than 6 indicates a depositing water, and more than 6 indicates corrosion. The higher the RI, the more corrosive is the water in the lines.

Phosphate Treatment

There are three types of corrosion-inhibiting phosphates: orthophosphates, polyphosphates, and zinc-containing phosphates. *Orthophosphates* are simple phosphate compounds: phosphoric acid (H_3PO_4), sodium phosphate (Na_3PO_4), sodium monohydrogen phosphate (Na_2HPO_4), and sodium dihydrogen phosphate

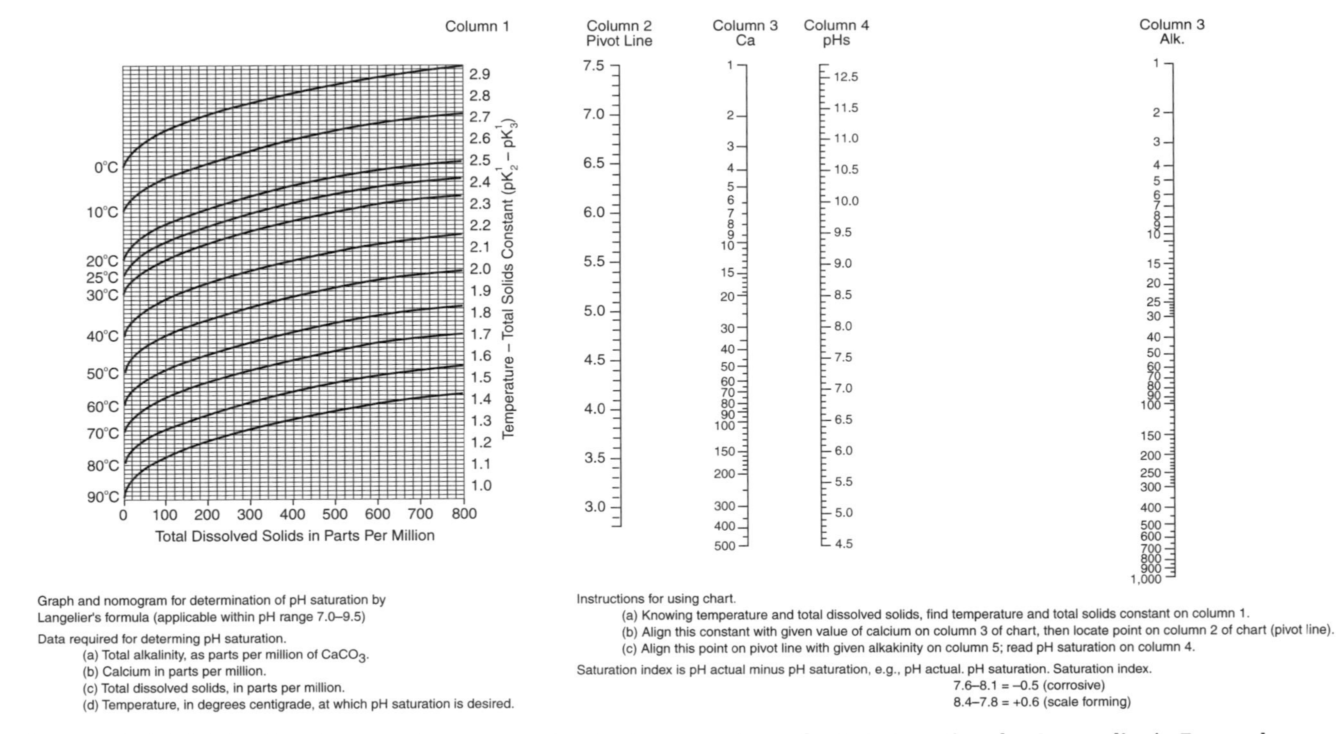

Graph and nomogram for determination of pH saturation by Langelier's formula (applicable within pH range 7.0–9.5)

Data required for determing pH saturation.

(a) Total alkalinity, as parts per million of $CaCO_3$.
(b) Calcium in parts per million.
(c) Total dissolved solids, in parts per million.
(d) Temperature, in degrees centigrade, at which pH saturation is desired.

Instructions for using chart.

(a) Knowing temperature and total dissolved solids, find temperature and total solids constant on column 1.
(b) Align this constant with given value of calcium on column 3 of chart, then locate point on column 2 of chart (pivot line).
(c) Align this point on pivot line with given alkakinity on column 5; read pH saturation on column 4.

Saturation index is pH actual minus pH saturation, e.g., pH actual. pH saturation. Saturation index.

7.6–8.1 = –0.5 (corrosive)
8.4–7.8 = +0.6 (scale forming)

Figure 14-2 Riehl's Graph and Nomogram for Determination of pH Saturation by Langelier's Formula

(NaH_2PO_4). *Polyphosphates* are long-chain phosphates formed by reacting phosphoric acid with sodium or potassium compounds. A common example of polyphosphate is sodium hexametaphosphate, $6(NaPO_3)$, which is commonly known as Calgon. Polyphosphates have been reported to be effective in reducing red water complaints. Their function differs under different conditions of water and their dose. At a low pH such as 5, they form a protective coating on the cathodic site in the presence of calcium and iron ions. At high pH and a low dose, they dissolve iron and calcium by a sequestering mechanism, thus preventing excessive scale formation.

Polyphosphates are more effective for corrosion control at high velocities of water and at high doses. They remove corrosion products from the anode by forming positively charged colloidal particles of ferric oxide and calcium compounds and metaphosphates, which are then deposited on the cathode area. Increasing polarization of the cathode reduces corrosion. If little or no calcium is present in the water, sodium hexametaphosphate will attack iron and cause corrosion. The calcium/polyphosphate ratio should be between 0.2 and 0.5.

Rusty water is not always an indication of corrosion, as iron may be naturally present in water and red scales may be a deposition of calcium together with oxidized natural iron.

Zinc-containing phosphates contain zinc in various concentrations (10–30 percent) with orthophosphates or polyphosphates. The protective film is formed of zinc phosphate and carbonates. The higher zinc concentration acts faster by rapid film formation. The higher the pH, the lower the zinc required for adequate control.

Questions

1. Corrosion is an electrochemical reaction. True or False.
2. Give the three parts of the corrosion process.
3. A corroding metal becomes anodic or cathodic.
4. Which are the common electron acceptors?
   a. Lead and copper
   b. Calcium ions
   c. Oxygen and hydrogen ions
5. In galvanic corrosion, a metal at the top in the electromotive series will serve as (a/an)
   a. Cathode
   b. Anode
6. Define polarization and depolarization.
7. What is a tubercle formed of?
8. What is a sequestering agent? Give an example.
9. What is the function of Calgon when used in very low doses?
10. Give the difference between Langelier and Ryzner indices. If the LSI is 0, is water stable, corrosive, or depositing?
11. Briefly give the use of the following in corrosion control:
    a. Sodium hexametaphosphate
    b. Zinc orthophosphate
12. In case of a depositing water, what gets deposited?
    a. Calcium carbonate
    b. Sodium carbonate

15 Disinfection

Disinfection of water or sewage means destruction or inactivation of waterborne pathogens (disease-causing microorganisms). Pathogens are present in our gastrointestinal tract and are discharged with fecal matter into the sewage system, which enters the sources of our water supply, especially surface waters. Therefore, water becomes their carrier.

At present, the well-established waterborne diseases are cholera, typhoid, paratyphoid, Legionnaire's disease, bacillary dysentery, amoebic dysentery, giardiasis, cryptosporidiosis, and infectious hepatitis. Out of these waterborne pathogens, *Giardia lamblia*, causing giardiasis, and *Cryptosporidium parvum*, causing cryptosporidiosis, are the hardest to kill. Waterborne pathogens exist in small numbers and are difficult to identify; therefore, coliform bacteria that are commonly and abundantly present in human excreta are used as indicator organisms of fecal contamination and thus a possible presence of waterborne pathogens. The destruction of the coliform group is a good criterion to determine adequate disinfection. Coliform bacteria are represented by fecal (mainly *Escherichia coli*) and nonfecal lactose fermenting, nonsporulating, gram negative (red stained), aerobic, and facultative bacilli (rod shaped).

The first large-scale disinfection of a public water supply in the United States was undertaken in 1908 at Jersey City, N.Y., by using chlorine. Chlorine has been a popular disinfectant in North America; it is effective, economical, and commonly available. The free residual chlorine, however, reacts with naturally occurring organics as humic acids and bromides to produce carcinogenic trihalomethanes (THMs). After this discovery in 1974, alternative disinfectants—such as chloramines, chlorine dioxide, ozone, potassium permanganate, hydrogen peroxide, silver, high pH, and ultraviolet radiation, alone or in combination—have been explored. *The goal is the proper balance between adequate disinfection and acceptable levels of disinfection by-products.* Of these disinfectants, chloramines, chlorine dioxide, and ozone are the most considered and well-studied practical alternatives.

Chlorine as a Water Disinfectant

Chlorination is the act of applying chlorine to water. It is the most common method of disinfection. Before 1974, more than 95 percent of municipalities used chlorine for disinfection. Besides disinfection, chlorine removes tastes, odors, iron, and manganese and controls slime-producing bacteria. It also helps in coagulation and destruction of cyanides and phenols.

Chlorine Properties

Discoverer Carl Wilhelm Scheele first isolated chlorine in 1774. At room temperature, chlorine is a greenish yellow gas. Its boiling point is –34.6°C and its melting point is –101°C. It can, therefore, be compressed easily into a liquid state. One mL of liquid forms 450 mL of gas. The density of chlorine is 3.21 g/L at standard temperature and pressure, while air is only 1.29 g/L. Thus, chlorine is about 2.5 times heavier than air. It is a very strong oxidizing agent. It reacts with various types of organic substances, metals, and ammonia. Chlorine has a strong, unpleasant smell and is corrosive to metals and flesh when wet.

Chlorine reacts with water to produce hydrochloric acid and hypochlorous acid. Hypochlorous acid (HOCl) is a strong disinfectant.

$$H_2O + Cl_2 \rightarrow HCl + HOCl$$

Hypochlorous acid is a weak acid. It ionizes into hypochlorite ion (OCl^-) and hydrogen ion (H^+) in water; the latter forms hydronium ion (H_3O^+).

$$HOCl + H_2O \rightarrow H_3O^+ + OCl^-$$

According to the old theory, hypochlorous acid decomposes into hydrochloric acid (HCl) and nascent oxygen (O by itself); the latter combines with the oxidizable cell structures in the microorganisms and kills them. Recent theory states that chlorine unites with the enzymes of microorganisms (possibly at the nitrogen site) and inactivates them. Enzymes are proteins, which are formed of four essential elements: nitrogen, carbon, hydrogen, and oxygen.

Sometimes, chlorination causes the disappearance of the bacterial bodies, probably by converting them into soluble compounds. Whatever the mechanism, chlorine inactivates the pathogens and renders the water potable or safe for drinking.

Various Forms of Chlorine Used in Disinfection

Pure Chlorine

Pure chlorine is available in a gas or liquid state in 100-, 150-, or 2,000-lb containers. It is also available in tank car lots of 15–90 tons. It is applied to the water by means of a special apparatus for dosage control known as a chlorinator.

Hypochlorites

Hypochlorites of sodium and calcium are solid forms of chlorine. They produce hypochlorous acid when dissolved in water. High-test hypochlorite, commonly known as *HTH,* is calcium hypochlorite ($Ca(ClO)_2$), which contains about 70 percent available chlorine. Sodium hypochlorite (NaClO) is only 15 percent available chlorine and is common household bleach.

$$Ca(ClO)_2 + 2H_2O \rightarrow Ca(OH)_2 + 2HOCl$$

Hypochlorites are used as solutions in water disinfection. The apparatus used to dispense them into the water is known as a *hypochlorinator.*

Chloride of Lime

Chloride of lime is also known as bleaching powder. Its chemical formula is CaClOCl. It is produced by passing chlorine gas over slaked lime ($Ca(OH)_2$).

$$Ca(OH)_2 + Cl_2 \rightarrow CaClOCl + H_2O$$

Chloride of lime contains 33–35 percent available chlorine. It is also applied as solution. Like hypochlorites, it produces hypochlorous acid in water.

$$2CaClOCl + 2H_2O \rightarrow CaCl_2 + Ca(OH)_2 + 2HOCl$$

Breakpoint Chlorination

Breakpoint chlorination is a technique used in chlorination of waters containing organic matter and ammonia to ensure proper disinfection by producing free residual chlorine. Increasing doses of chlorine are added to a series of samples.

At first, residual chlorine increases to a point, known as a "hump," after which a further increase in the dose results in a decrease of residual chlorine, which reaches the lowest point, the "breakpoint" or "dip." After the

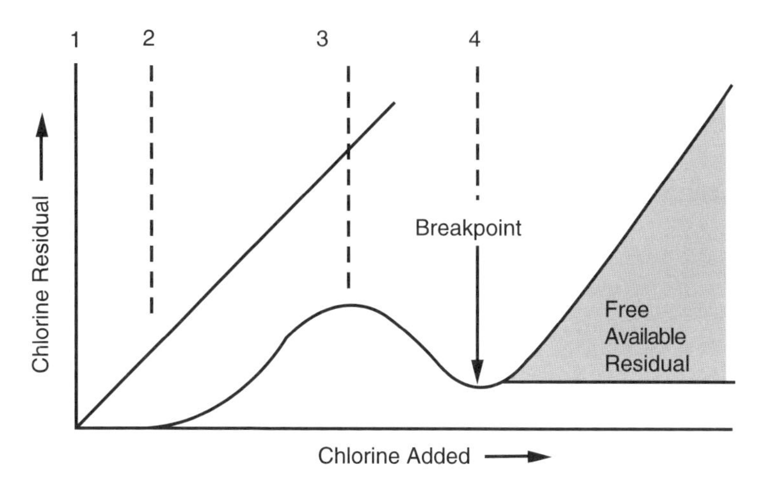

NOTE: Numbers indicate points within treatment where chlorine is added.

Figure 15-1 Breakpoint Chlorination Curve

breakpoint, an increase in dose results in a corresponding increase in residual chlorine. This residual chlorine is known as *free residual chlorine* or *breakpoint residual,* which is hypochlorous acid under normal pH (below 7.5) conditions. The breakpoint indicates the completion of reactions of chlorine with bacteria and all other substances such as iron (Fe), manganese (Mn), hydrogen sulfide (H_2S), and ammonia (NH_3). Breakpoint chlorination affords, in many cases, better control of bacteria, tastes, odors, and red and black waters. Figure 15-1 illustrates the breakpoint chlorination curve.

Chemistry of Breakpoint Chlorination

The sequence in which chlorine reacts with various substances in the water is as follows:

1. *Reducing agents:* These agents include iron, manganese, hydrogen sulfide, nitrites, and thiosulfates, which neutralize chlorine and result in the formation

of chlorides. Thus, iron, manganese, and hydrogen sulfide are removed. They cause red water, black water, and tastes and odors, respectively. Due to neutralization of chlorine into chlorides, there is no residual chlorine at this stage. Chlorides do not have any disinfecting properties.

$H_2S + 4Cl_2 + 4H_2O \rightarrow H_2SO_4 + 8HCl$ at high pH

$H_2S + Cl_2 \rightarrow 2HCl + S$ at low pH

2. *Organic matter:* Organic substances, including microorganisms and phenols, react with chlorine after reducing agents to produce chloroorganic compounds. This area in the curve also presents taste-and-odor problems.
3. *Ammonia:* Chlorine forms chloramines (NH_2Cl) with ammonia until a hump is reached.

   The following reactions occur:

$$NH_3 + HOCl \rightarrow NH_2Cl + H_2O$$

   Hypochlorous acid is the form of chlorine produced after reacting with water; therefore, instead of Cl_2, HOCl is used in the equations. Both chloroorganic compounds and chloramines are known as *combined residual chlorine.* If chlorination comes after sedimentation and filtration, the combined residual chlorine is commonly represented by chloramines due to the presence of ammonia and the absence of organic matter.
4. *Destruction of combined residual chlorine:* Chlorine reacts with chloroorganic compounds and chloramines and thus gets neutralized, resulting in a drop of residual chlorine. The end products of these reactions are nitrogen gas (N_2), nitrous oxide (N_2O), hydrochloric acid, and nitrogen trichloride (NCl_3). Nitrogen trichloride is also known as trichloramine. These reactions occur between the hump and the breakpoint. After the breakpoint (dip), chlorine in the water is free residual chlorine mostly as hypochlorous acid.

The object of breakpoint chlorination is to produce and maintain free residual chlorine because free residual chlorine can exist in the water only after the breakpoint or complete oxidation of substances, which react with chlorine.

Chlorine Demand

Chlorine demand is the amount of chlorine consumed in the water during a certain contact period, commonly 30 min. It is calculated by subtracting residual chlorine from the chlorine dose. Suppose we used a 5-mg/L chlorine dose in a sample. After 30 min (normal contact time period), there is 2 mg/L residual. The chlorine demand of water is 3 mg/L.

Factors Affecting Chlorination

The effectiveness of chlorine is affected by pH, temperature, contact time period, concentration, and residual chlorine type. The amount of hypochlorous acid depends on the pH of the water. The following table shows the predominant form of free residual chlorine at various pH values:

pH	Predominant Free Residual Chlorine
6–7.5	HOCl, chlorine as hypochlorous acid
7.5–9	OCl^-, chlorine as hypochlorite ion

Hypochlorous acid is ionized more than 50 percent into H^+ and OCl^+ ions above pH 7.5. Since hypochlorous acid and not the hypochlorite ion (OCl^-) is the principal disinfectant, effectiveness of chlorine starts declining above pH 7.5. The following table suggests the minimum free residual chlorine required at different pH values for disinfection:

pH	Free Residual Chlorine, *ppm*
6–8	0.2
8–9	0.4
9–10	0.8

These amounts at normal water temperature (about 20°C) will disinfect water adequately in about 10 min contact time.

Effectiveness of chlorine varies directly with temperature. The higher the temperature, the quicker the kill of microorganisms; thus, less contact time is required. Chlorine requires a certain amount of time to react with microorganism cells. The longer the contact time, the more effective is the disinfection. That is why the CT concept is adopted for proper disinfection. C stands for *concentration of disinfectant* as milligrams per litre and T stands for contact time in minutes. CT is a constant for each disinfectant. The contact period normally recommended for chlorine is 15–40 min. Thirty minutes are allowed for proper disinfection.

Concentration of free residual chlorine under different pH and temperature conditions is an important factor. Minimum effective amounts of free residual chlorine at different pH values are already discussed. Normally, 0.5–1 mg/L free residual chlorine will effectively disinfect the water; remove iron, manganese, tastes, odors, and colors; and control slime and algae.

Combined residual chlorine (chloramines) is less effective than free residual chlorine (HOCl). Normally, chloramines require a higher dose and longer contact time than free residual chlorine for the same degree of disinfection.

Types of Chlorination

Prechlorination may be defined as the application of chlorine to raw water, which means before any water treatment. It helps in the coagulation and removal of tastes and odors. It kills a large number of bacteria, algae, and other microorganisms, which settle out in the clarifiers. Prechlorination is especially useful for surface water treatment and ensures disinfection while water flows through pipes and the treatment plant. The chlorine residual should be 0.2–0.5 mg/L in the settled water reaching the filters; a higher amount can cause formation of THM and other harmful disinfection by-products. The chemical nature of these chemicals is discussed in chapter 12.

Superchlorination is the application of a very high concentration of chlorine, primarily for sterilizing pipes and controlling tastes and odors caused by microorganisms. Normally, dechlorination is required after superchlorination.

Postchlorination is the chlorination of completely treated water. It is used for proper disinfection and to maintain the desired amount of residual chlorine in the reservoir and distribution system as a way to prevent any accidental contamination and slime-forming bacterial growth. The presence of residual chlorine in the system indicates that an adequate amount of chlorine has been used for disinfection. The amount of residual chlorine is a quicker method than bacteriological tests to determine potability of water.

Dechlorination is the destruction of chlorine by using activated carbon or reducing agents, such as sulfur dioxide (SO_2), sodium sulfite (Na_2SO_3), sodium bisulfite ($NaHSO_3$), or sodium thiosulfate ($Na_2S_2O_3$). Dechlorination is required to maintain the desired level of residual chlorine in the water. It is mainly done after superchlorination or in water supplies for some industrial uses. Chlorine is not desirable for some ion exchange resins, which are used in water softeners.

Hypochlorination is the use of hypochlorites for chlorination. Hypochlorites of sodium and calcium are common compounds used for this purpose. They are available in powder or tablet form; thus, they have the advantage of easy storage and application. They are especially useful for disinfecting lines and discontinuous treatment of small-scale operations. Hypochlorites react with water and produce hypochlorous acid and hydroxides. Hydroxides tend to raise the pH, which may cause the precipitation of calcium carbonate ($CaCO_3$). A complex of phosphate, sodium hexametaphosphate, is added to overcome this problem.

Hypochlorites are used as solutions by means of hypochlorinators. All hypochlorites are corrosive to some degree; therefore, store them in wood, glass, plastic, or rubber containers.

Chlorine and Health

Chlorine is a highly toxic chemical even in small concentrations in air. The following table shows the physiological effects of various concentrations of chlorine by volume in air:

Effects	Chlorine, *ppm in air*
Least detectable by odor	Below 3.5
Produces throat irritation	15
Produces coughing	30
Dangerous after 30-min exposure	40–60
Rapidly fatal	1,000

Twenty to 50 mg/L of chlorine in water for a few days did not show any bad physiological effects.

Chlorine and Safety

When using chlorine, observe the following precautions:

1. Use a mask when entering a chlorine-containing atmosphere.
2. Apparatus, lines, and cylinder valves should be checked regularly for leaks. Use ammonia fumes to test leaks. Ammonia and chlorine produce white fumes of ammonium chloride, which indicate leaks.
3. Because it is heavier than air, always store chlorine on the lowest floor; it will collect at the lower level. For the same reason, never stoop down when a chlorine smell is noticed.

Handle chlorine carefully and respectfully, as she is the "green goddess of water."

Chloramines as Water Disinfectants

Chloramines are produced by reacting ammonia with chlorine. Despite their weaker disinfecting properties and slower disinfection rate, they have been used successfully by a large number of utilities in the United States. Ammonia is used ahead of chlorine to prevent formation of THMs. Reaction rate and type of chloramine, such as mono-, di-, or trichloramine formation, depend on the pH and chlorine-to-ammonia ratio. At pH 8 and a ratio of chlorine to ammonia between 4 and 5, monochloramine is the predominant chloramine species, which is preferred because it is less odorous. At a pH of 5–8, both mono- and dichloramines are present; at pH 4.4–5, dichloramine predominates; and below pH 4.4, trichloramine predominates. As the chlorine-to-ammonia ratio progresses, monochloramine (NH_2Cl)

changes to dichloramine ($NHCl_2$), then to trichloramine (NCl_3), and finally to nitrogen and chlorides until free residual chlorine is formed.

$$NH_3 + HOCl \rightarrow NH_2Cl + H_2O$$

$$NH_2Cl + HOCl \rightarrow NHCl_2 + H_2O$$

$$NHCl_2 + HOCl \rightarrow NCl_3 + H_2O$$

$$2NH_2Cl + HOCl \rightarrow 3HCl + H_2O + N_2\uparrow$$

Chloramines are more effective at a higher pH (above 8) and require about 70 times more contact time and much higher concentration than hypochlorous acid. Chloramines, however, are comparable to hypochlorite ion, which is free residual chlorine above pH 7.5.

Some other disadvantages of chloramines are that they induce hemolytic anemia and thus cause problems with the use of dialysis machines. They are harmful for tropical fish. Due to these disadvantages, they are not accepted as good disinfectants by many utilities.

Bromine as a Water Disinfectant

Bromine is a halogen with disinfection properties like chlorine. It is a dark red liquid about three times heavier than water. It readily evaporates, and its vapor is very irritating to the eyes and throat. It forms in water hypobromous acid and hydrobromic acid. Hypobromous acid is a strong disinfectant.

$$H_2O + Br_2 \rightarrow HBr + HOBr$$

The major difference between hypobromous acid and hypochlorous acid is that its effectiveness starts declining at pH 8.5. Bromine produces bromoamines with ammonia, which are as effective as hypobromous acid. Bromine has been used for disinfection of swimming pools and waters with high ammonia content.

Because bromine is scarce, it is more expensive than chlorine. It also is physiologically more active to humans; thus, its use as a disinfectant is limited.

Iodine as a Water Disinfectant

Iodine (I_2), a bluish black solid, is another halogen with disinfecting properties. Iodine forms hypoiodous (HOI) acid in water, which is an effective disinfectant.

$$I_2 + H_2O \rightarrow HI + HOI$$

Iodine's great advantage is that it does not react with ammonia. Its effectiveness is not affected by pH. By appropriate control of the dose, it provides a broad spectrum of germicidal capability. Iodine is quite expensive, scarce, and physiologically active, which results in limited use. It is useful in a small-scale, noncontinuous application, such as swimming pool disinfection and as tablets for individual water supply disinfection in the armed forces.

Chlorine Dioxide as a Water Disinfectant

Sir Humphry Davi first discovered chlorine dioxide, a yellow to red gas, in 1811, by reacting potassium chlorate with hydrochloric acid.

$$4KClO_3 + 4HCl \rightarrow 4KCl + 4ClO_2\uparrow + 2H_2O + O_2\uparrow$$

It was first used by a water utility in Niagara Falls in 1944 for taste and odor control. By 1997, about 400 water utilities in North America used it for predisinfection. Commercially, it is produced by reacting sodium chlorite with chlorine and/or an acid.

$$2NaClO_2 + Cl_2 \rightleftarrows 2NaCl + 2ClO_2\uparrow$$

$$5NaClO_2 + 4HCl \rightleftarrows 5NaCl + 4ClO_2\uparrow + 2H_2O$$

Both processes use a similar generator formed of a cylindrical reactor of polyvinyl chloride or Pyrex glass. The Rio-Linda in Sacramento, Calif., generator uses chlorine gas rather than chlorine solution and the yield is improved to 96–98 percent compared to 95 percent or less in other generators where pH is regulated to 3–4 by adjusting chlorine feed for a better efficiency.

Electrolysis of $NaClO_2$ is another method, which is quite promising, as it provides 100 percent chlorine dioxide solution. There is oxidation of ClO^{-}_2 to ClO_2 at the anode and reduction of Na^+ to Na at the cathode.

$$\underset{}{NaClO_2} \rightarrow \underset{\text{Cathode}}{Na} + \underset{\text{Anode}}{ClO_2\uparrow}$$

Chlorine dioxide is a strong disinfectant and, unlike chlorine, its effectiveness is not affected by ammonia and pH; furthermore, it does not produce THMs. Its effectiveness against *Giardia* and *Cryptosporidium* is quite good. It is very promising when chlorine dioxide treatment is followed by chlorine or chloramines. This sequential disinfection has a synergistic effect. A 1.5-mg/L chlorine dioxide dose gave more than 90 percent inactivation; whereas, this dose followed by 1.6 mg/L chlorine or 2.8 mg/L chloramines gave about 99 percent inactivation of these pathogens. Chloramine and chlorine alone, however, had insignificant effects. Chlorine dioxide apparently weakens the organisms to be effectively destroyed by chlorine or chloramines.

There are disadvantages of using chlorine dioxide. It is relatively expensive to generate and is explosive at a concentration above 10 percent in the air. Chlorine dioxide is unstable and reverts to chlorite, which can cause anemia in some individuals. Therefore, it is generated at the site and applied immediately. Also, sodium chlorite is hazardous to handle. Chlorine dioxide use is further limited due to the potential health effects of chlorite and chlorate, the by-products. Thus, the use of chlorine dioxide in the water industry is somewhat limited mainly due to chlorite formation and its dose restriction.

Various chemicals such as activated carbon, sulfur dioxide, sulfite, and ferrous compounds have been studied to control chlorite. It is well documented that chlorite can be effectively controlled by using ferrous ions (Fe^{+2}) in water treatment. About 3 mg/L ferrous ion reduces chlorite by 1 mg/L, which is a stoichiometric ratio.

Ozone as a Water Disinfectant

Ozone, a colorless gas with a peculiar pungent odor, is a triatomic form of oxygen (O_3). It was discovered by M. van Marum in 1785 and was named ozone by C.F. Schönbein in 1840 after the Greek word *ozein*, meaning *to smell*. It is generated by passing air or oxygen through a high-voltage arc between two electrodes, or ultraviolet radiation. Commercially, it is generated by a silent electric discharge or corona discharge by using a high-voltage (over 15,000 volts). A small percentage of oxygen is converted (1–3 percent from air and 2–6 percent from oxygen feed) into ozone in this process. It is 20 times more soluble in water than oxygen, which helps in its dispersion in water.

$$\underset{\text{Oxygen}}{3O_2} \rightleftarrows \underset{\text{Ozone}}{2O_3}$$

Ozone is a very unstable gas; its molecule readily decomposes into an oxygen molecule and nascent oxygen. Its half-life at room temperature is only 15–20 min. Nascent oxygen apparently causes disinfection.

$$O_3 \rightarrow O_2 + O$$

Immediately after generation, the ozone is dispersed through the water. Microorganisms are killed immediately upon contact with the ozone. Their cells are ruptured (lysis), resulting in death. Ozone is the strongest disinfectant used in water treatment. Ozone has very strong germicidal properties with very rapid effects. Its effectiveness is not impaired by ammonia and pH. It leaves dissolved oxygen after its decomposition. Ozone's disadvantages are that it must be generated at the site of application, does not have a residual effect, and is quite expensive to generate. Due to these shortcomings, ozonation is not as common as chlorination. It is more accepted in France, Switzerland, and Germany (over 1,000 installations). They believe in treating the water to pure, unpolluted groundwater form without any chemical odor. The first major ozone installation was in 1905 in Nice, France, followed by Paris in 1906. Canada has about 20 major installations, including a 60-mgd plant in Quebec. In the United States, the first major installation was at Whiting, Ind., in 1939, followed by Philadelphia in 1949.

The ozonation treatment of water can be divided into three parts: preparation of feed gas, production, and contaction. *Preparation* of feed gas is mainly filtration of air to remove dust and then drying, as moisture produces nitric acid from air feed by combining oxygen and nitrogen. Oxygen feed is a more efficient system for better conversion rate and less corrosion problems due to lack of nitrogen. *Production* is simply passing the feed at low pressure between two electrodes separated by a dielectric and a gap across which an alternating potential of about 15,000 volts is maintained for ozone production. A large amount of heat is released in the process, which needs a cooling system, primarily using water. *Contaction* is the passing of the ozone oxygen/air mixture through the water by dispersing it mostly by diffusers at the bottom of a contact chamber to have the maximum transfer of ozone into the water.

Ozone, besides being an excellent disinfectant, controls tastes and odors, color, algae, slime growth, THM formation and other organics, cyanides, sulfides, sulfites, iron, manganese, and turbidity. There are some disadvantages, such as the high cost of production; there is almost no residual effect; and it is difficult to adjust to the variations in treatment, load, or demand. It is also selective in oxidation of some organics such as ethanol, which does not readily react with it and sometimes causes fragmentation of large molecules that encourages bacterial growth in the distribution system. Ozonation is followed by chlorination or chloramination for the residual effect. Furthermore, its by-products are still not properly understood.

Due to high capital cost and lack of residual effect, various combinations of ozone and other disinfectants have been studied.

Peroxone is the use of hydrogen peroxide and ozone together. It is demonstrated that this combination accelerated the oxidation of some organics 2–6 times. Peroxone is known as an advanced oxidation process, which produces hydroxyl (OH^-) radical, a very active intermediate that generally has far greater oxidizing power than ozone by itself. Hydrogen peroxide and ozone in a ratio of 2:4 mg/L are very effective. Peroxone is almost one half as expensive as ozone alone. Peroxone (2:4 mg/L) followed by 1.5 mg/L of chloramines has been an effective disinfection treatment for some utilities.

Soozone, a combination of ultrasonic waves and ozone, has been used in Indian Town, Fla., to treat sewage. Ultrasonic waves break organic particles and ozone oxidizes them.

Ultraviolet Light

Ultraviolet light rays cause death of microorganisms by oxidation of their enzymes. The most effective wavelength is 2,650 angstrom (1 Å = 10^{-8}cm). Thus, rays with a wavelength less than 3,100 Å are effective. The mercury vapor lamp is an economical method of producing ultraviolet light of 2,537 Å. For disinfection with ultraviolet light, water should be clear, colorless, and shallow (3–5 in. deep) to allow effective penetration of rays. These requirements as well as no residual effect and cost of application limit the use of this method to very small water supplies.

Silver Ions

Silver ions in very low concentration (0.01 ppm) are sometimes used for water disinfection. Disinfection is obtained by passing water between electrically charged silver plates, which disperse silver ions into the water. The effectiveness of silver ions is decreased by the presence of organic substances in water. Silver is not a practical disinfectant for water supplies. It is too expensive and has a very limited use.

Questions

1. What is the main purpose of disinfection of water?
2. Why is chlorine used commonly for water disinfection?
3. Are there any other uses of chlorination besides disinfection?
4. Disease-producing bacteria are called:
   a. Aerobic
   b. Facultative
   c. Pathogenic
5. Chlorine is a highly toxic gas. True or False.
6. Chlorine is lighter than air. True or False.
7. Free residual chlorine is a better disinfectant than combined residual chlorine. True or False.

8. Should chlorine be stored on the upper or lower floor of the building?
9. Postchlorination is the chlorination of water before any treatment. True or False.
10. How would you detect a chlorine leak?
    a. By feeling a leak.
    b. By looking for a greenish gas coming out.
    c. By using carbon dioxide vapor.
    d. By using ammonia vapor.
11. Name six waterborne diseases.
    (1) ______________________________
    (2) ______________________________
    (3) ______________________________
    (4) ______________________________
    (5) ______________________________
    (6) ______________________________
12. What are the characteristics of coliform bacteria?
13. How does chlorine kill microorganisms?
14. Define prechlorination, postchlorination, and hypochlorination.
15. Differentiate between free residual and combined residual chlorine.
16. What is the purpose of breakpoint chlorination?
17. Draw a breakpoint curve and explain its various parts.
18. Which of these is a preferred chloramine species and why?
    a. Monochloramine
    b. Dichloramine
    c. Trichloramine
19. What is the ratio of chlorine to ammonia for proper chloramination?

20. Chlorine dioxide is one of the preferred predisinfectants. Why?
21. What is the end product of chlorine dioxide and how is it controlled?
22. Give advantages and disadvantages of ozonation.
23. Ozone is a form of oxygen. True or False.
24. Chlorine or chloramines are more effective when used after ozone or chlorine dioxide treatment. True or False.
25. CT value is the product of milligrams pre litre of the disinfectant and its contact time in minutes. True or False.

Appendix

Chemical Name	Common Name	Chemical Formula	Used for
Aluminum sulfate	Alum	$Al_2(SO_4)_3 \cdot 14H_2O$	Coagulation
Ammonia	Ammonia gas	NH_3 (ammonia gas)	
	Ammonia aqua	NH_4OH (ammonia solution)	Chloramination
Calcium bicarbonate		$Ca(HCO_3)_2$	Alkalinity
Calcium carbonate	Limestone	$CaCO_3$	
Calcium hydroxide	Hydrated lime or slaked lime	$Ca(OH)_2$	Softening
Calcium hypochlorite	HTH	$Ca(ClO)_2$	Chlorination
Calcium oxide	Unslaked lime or quick lime	CaO	Softening
Carbon	Activated carbon	C	Taste, odors, and pesticide removal
Chlorine		Cl_2	Disinfection
Chlorine dioxide		ClO_2	Disinfection
Copper sulfate	Blue vitriol	$CuSO_4 \cdot 5H_2O$	Algae control

Some Chemicals Used in Water and Wastewater Treatment

Table continues next page

Chemical Name	Common Name	Chemical Formula	Used for
Ferric chloride		$FeCl_3 \cdot 6H_2O$	Coagulation
Ferric sulfate		$Fe_2(SO_4)_3$	Coagulation
Ferrous chloride		$FeCl_2$	Chlorite control
Fluosilicic acid (hydrofluosilicic acid)	Fluoride	H_2SiF_6	Fluoridation
Hydrochloric acid	Muriatic acid	HCl	
Ozone		O_3	Disinfection
Potassium dichromate		$K_2Cr_2O_7$	
Potassium permanganate		$KMnO_4$	Taste and odor control
Sodium aluminate		$NaAlO_2$	Coagulation
Sodium bicarbonate	Baking soda	$NaHCO_3$	Alkalinity
Sodium carbonate	Soda ash	Na_2CO_3	Softening
Sodium chloride	Salt	$NaCl$	

Some Chemicals Used in Water and Wastewater Treatment (continued)

Table continues next page

Chemical Name	Common Name	Chemical Formula	Used for
Sodium chlorite		$NaClO_2$	Chlorine dioxide formation
Sodium fluoride		NaF	Fluoridation
Sodium fluosilicate		Na_2SiF_6	Fluoridation
Sodium hexametaphosphate	Calgon	$Na_6(PO_3)_6$ or $6(NaPO_3)$	Sequestering
Sodium hydroxide	Lye	$NaOH$	Alkalinity
Sodium hypochlorite	Bleach	$NaClO$	Chlorination
Sodium phosphate		$Na_3PO_4 \cdot 12H_2O$	
Sodium thiosulfate		$Na_2S_2O_3$	
Sulfuric acid	Oil of vitriol	H_2SO_4	
Zinc orthophosphate		$Zn_3(PO_4)_2$	Corrosion control

Some Chemicals Used in Water and Wastewater Treatment (continued)

Some Equations Common in Water and Wastewater Chemistry

Softening

$$Ca(OH)_2 + Ca(HCO_3)_2 \rightarrow 2CaCO_3\downarrow + 2H_2O$$

$$Ca(OH)_2 + Mg(HCO_3)_2 \rightarrow CaCO_3\downarrow + MgCO_3 + 2H_2O$$

$$Ca(OH)_2 + MgCO_3 \rightarrow CaCO_3\downarrow + Mg(OH)_2\downarrow$$

$$CaSO_4 + Na_2CO_3 \rightarrow CaCO_3\downarrow + Na_2SO_4$$

$$Ca(OH)_2 + MgSO_4 \rightarrow CaSO_4 + Mg(OH)_2\downarrow$$

Coagulation

$$Al_2(SO_4)_3 + 3Ca(OH)_2 \rightarrow 2Al(OH)_3\downarrow + 3CaSO_4$$

$$Fe_2(SO_4)_3 + 3Ca(OH)_2 \rightarrow 2Fe(OH)_3\downarrow + 3CaSO_4$$

Disinfection

$$Cl_2 + H_2O \rightarrow HCl + HOCl$$

$$Ca(OCl)_2 + 2H_2O \rightarrow Ca(OH)_2 + 2HOCl$$

$$NH_3 + HOCl \rightarrow NH_2Cl + H_2O$$

$$NH_3 + 2HOCl \rightarrow NHCl_2 + 2H_2O$$

Neutralization

$$CO_2 + Ca(OH)_2 \rightarrow CaCO_3\downarrow + H_2O$$

$$CaCO_3 + H_2SO_4 \rightarrow CaSO_4 + H_2O + CO_2\uparrow$$

$$Ca(HCO_3)_2 + H_2SO_4 \rightarrow CaSO_4 + 2H_2O + 2CO_2\uparrow$$

Compounds Causing Acidity and Alkalinity in Water

$$CO_2 + H_2O \rightarrow H_2CO_3$$

$$Al_2(SO_4)_3 + 3H_2O \rightarrow 2Al(OH)_3\downarrow + 3H_2SO_4$$

$$CaO + H_2O \rightarrow Ca(OH)_2$$

Glossary

Acid A compound that forms hydronium ions in water solution. It is a proton donor.

Acid anhydride A nonmetallic oxide that reacts with water to form an acid.

Adsorption The acquisition of a gas, liquid, or solid on the surface of a solid particle.

Alcohol An organic compound containing a hydrocarbon group and one or more –OH groups.

Allotropy The existence of an element in two or more forms in the same physical state.

Amphiprotic Capable of acting either as an acid or as a base.

Angstrom A unit of linear measure; 10^{-8} cm.

Anhydrous Without water of crystallization.

Anion A negative ion.

Anode Electrode that attracts anions. A positively charged pole. The electrode where oxidation takes place.

Atom The smallest particle or unit of an element that is capable of entering into combinations with other elements.

Atomic number The number of protons in the nucleus of an atom.

Atomic weight The relative average atomic mass of an element compared to $1/12$ the mass of carbon-12 isotope.

Avogadro number The number of carbon-12 atoms in exactly 12 g of this isotope: 6.022169×10^{23}.

Base A substance that accepts protons from another substance.

Base, conjugate The part of an acid molecule that is left after it has donated protons.

Basic anhydride A metallic oxide that combines with water to form a hydroxide.

Binary compounds Compounds that are formed of two elements.

Blue vitriol Hydrate copper (II) sulfate, $CuSO_4 \cdot 5H_2O$.

Boiling point The temperature at which equilibrium vapor pressure is equal to the prevailing atmospheric pressure.

Brownian motions The peculiar dancing movements of colloid particles as a result of collisions with the molecules of water.

Buffer A substance that resists any change in pH.

Calorie The quantity of heat required to raise the temperature of 1 g of water by 1°C.

Catalyst A substance that alters the rate of a chemical reaction but remains unchanged itself at the end of the reaction.

Cathode A negatively charged electrode. It attracts cations. The electrode where reduction occurs.

Cation A positively charged ion.

Chemical bond The bondage between atoms produced by transfer or sharing of electrons.

Chemical change A change in which new chemicals with new properties are formed.

Colloidal suspension Suspended particles (1–100 nm in diameter) in a dispersing medium.

Compound A substance composed of two or more elements combined in a definite proportion.

Concentrated Containing a large amount of solute.

Covalence Covalent bonding.

Covalent bond Bond formed by a pair of shared electrons.

Dehydration The removal of water from a substance.

Deliquescence Taking up water by a substance from the air to form a solution.

Density The mass per unit volume of a substance.

Diatomic Particle consisting of two atoms.

Diffusion The process of spreading out of particles to uniformly fill a space.

Dilute Containing a small amount of dissolved solute.

Dipole A polar molecule with one region positive and the other negative.

Diprotic An acid that can donate two protons per molecule.

Dissociation The separation of ions of an ionic solute during the solution formation.

Ductile Capable of being drawn into a wire.

Effervescence The rapid release of a dissolved gas from a liquid.

Electrochemical Pertaining to spontaneous oxidation reduction reactions used as a source of electric energy.

Electrolysis Decomposition of a substance by electricity.

Electrolyte A substance whose water solution conducts electricity.

Electron A negatively charged particle revolving around the nucleus of an atom.

Electronegativity The force by which shared electrons are attracted by a combining atom.

Electrovalence Ionic bonding.

Element A substance that cannot be decomposed by ordinary chemical means.

Emulsoid or emulsion Dispersion of a liquid in a liquid.

End point or equivalence point Point in a titration at which quantities of the standard and standardized chemicals are chemically equivalent.

Endothermic A reaction in which heat is absorbed.

Energy level A region around the nucleus of an atom in which electrons revolve.

Enzyme A catalyst produced by living cells.

Equilibrium A dynamic state in which two opposing processes proceed at the same rate at the same time.

Equilibrium vapor pressure The pressure of a vapor in equilibrium with its liquid.

Ester An organic oxide.

Evaporation The escape of molecules from the surface of liquids and solids.

Exothermic reaction A chemical reaction that liberates heat.

Fat An organic oxide formed of glycerol and long carbon chain acids.

Formality See *molarity*.

Formula A shorthand representation of the composition of a chemical using chemical symbols and numerical subscripts.

Freezing point The temperature at which a liquid becomes a solid.

Functional group A group of atoms occurring in many molecules that possesses a characteristic reactivity.

Gas A state of matter in which a substance does not possess a definite shape or volume.

Gram A metric unit of mass equal to the mass of 1 mL of water at 4°C.

Gram atomic weight The mass in grams of one mole of naturally occurring atoms of an element. It is the atomic weight of an element expressed in grams.

Heat A form of energy.

Heterogeneous Having different properties in different parts.

HTH, high-test hypochlorite Calcium hypochlorite, a solid form of chlorine.

Homogeneous Having uniform properties throughout.

Hydrate A crystallized substance with water of crystallization.

Hydration Association of water molecules to particles of the solute.

Hydrocarbon A compound composed of hydrogen and carbon.

Hydrogen bond A weak chemical linkage between a hydrogen atom in one polar molecule and the more electronegative atom in a second polar molecule of the same substance.

Hydrolysis A chemical reaction in which water is involved for decomposition.

Hydronium ion A hydrated proton (hydrogen ion); H_3O^+ ion.

Hygroscopic substance A substance that absorbs and retains moisture from the atmosphere.

Indicator A substance that changes in color from a standard reagent to the standardized.

Insoluble substance A sparingly soluble substance with the solubility less than 0.1 g per 100 g of water.

Ion An atom or group of atoms with an electric charge.

Ionic bonding Chemical bonding in which electrons are transferred from one atom to another.

Ionization The formation of ions from polar solute molecules by the action of polar molecules of the solvent.

Isomers Compounds having the same molecular formula but different molecular structure.

Kelvin (absolute scale) temperature Temperature on Kelvin scale where 273 = 0°C.

Kinetic energy Energy of motion.

Lime Calcium oxide, CaO, also known as quicklime, and $Ca(OH)_2$, slaked lime.

Lipids Organic compounds composed of carbon, hydrogen, and oxygen; fats; and oils.

Liquid A state of matter that has a definite volume but no definite shape.

Litre Volume occupied by 1 kg of water at 4°C.

Lye A commercial grade of sodium hydroxide or potassium hydroxide.

Malleable Capable of being hammered into different shapes.

Mass Quantity of matter of a body.

Mass number The total number of protons and neutrons in an atom.

Matter Anything that occupies space and has mass.

Melting point The temperature at which a solid changes into a liquid.

Metre A metric unit of length equal to 39.37 in. (1/10,000,000 of the north polar quadrant of the Paris Meridian).

Mixture A material composed of two or more substances, each retaining its own properties.

Molality An expression of the number of moles of a solute per kilogram of solvent.

Molar volume The volume of 1 mol of any gas in litres at 0°C and 760 mm Hg pressure. It is 22.4 L.

Molarity An expression of the moles of solute per litre of solution. It is also called formality.

Mole Amount of a substance in grams that contains an Avogadro number of its particles. Practically, it is gram atomic weight for an element and gram formula weight of a compound.

Molecular formula A chemical formula representing the composition of a molecule of a substance.

Molecular weight The formula weight of a molecular substance.

Molecule The smallest stable and neutral unit of a substance.

Monoprotic An acid capable of donating one proton per molecule.

Neutralization The reaction between a base and an acid to produce a salt and water.

Neutron A neutral particle in the atomic nucleus with almost the same mass as that of a proton.

Nonelectrolyte A substance that does not conduct electricity in its water solution.

Normal solution A solution containing 1 g equivalent weight of a solute per litre of solution.

Normality An expression of the number of gram equivalent weights of a solute per litre of solution.

Nucleus The positively charged and centrally located part of an atom.

Octet Outer shell of an atom having eight electrons.

Organic Regarding carbon compounds and their derivatives.

Oxidation A chemical reaction in which electrons are lost.

Oxidation number Number of electrons of an element participating in compound formation. It is equal to the electrons in a neutral atom minus the electrons in a combined atom.

Oxidation-reduction reaction Any chemical reaction in which electrons are transferred.

Oxidizing agent A substance that gains electrons in a chemical reaction.

Period A horizontal row of elements in the periodic table.

Permanent hardness (noncarbonate hardness) Hardness in water due to the presence of sulfates, nitrates, and chlorides of calcium and magnesium.

Peroxone A combination of hydrogen peroxide and ozone for disinfection.

pH Minus logarithm of the hydronium ion concentration expressed as moles per litre.

Physical change A change in which the chemical composition of a substance remains unchanged.

Physical properties Properties that can be determined without changing chemical composition of a substance.

Plasma An ionized gas. The fourth state of matter.

Polar covalent bond A covalent bond in which there is an unequal sharing of electrons and thus an unequal distribution of a charge.

Polar molecules Molecules formed by polar covalent bonding.

Polarization Accumulation of hydrogen gas and hydroxide ions on the cathode in a corrosion process.

Potential energy Energy of position.

Precipitation The separation of an insoluble compound from a solution.

Protein An organic compound formed essentially of carbon, hydrogen, oxygen, and nitrogen.

Proton A positively charged particle in the nucleus.

Radical A group of covalently bonded atoms carrying a charge.

Redox Regarding reduction-oxidation reactions.

Reducing agent A substance that loses electrons in a chemical reaction.

Reduction Any reaction in which electrons are gained.

Reversible reaction A chemical reaction in which the products reform the reactants.

Salt An ionic compound formed of cations and conjugate base anions.

Saturated (1) A solution in which the rate of dissolving is equal to the rate of crystallization. (2) An organic compound with single covalent bonds in the carbon chain.

Shell A part of an atom around the nucleus of an atom in which electrons revolve.

Slaked lime Calcium hydroxide, $Ca(OH)_2$.

Soft water Water with low calcium and magnesium contents.

Solute The dissolved substance in a solution.

Solution The homogeneous mixture of two or more substances.

Solution equilibrium The physical state in a solution at which the rate of crystallization is equal to the rate of dissolving.

Solvent A dissolving medium in a solution.

Soozone A combination of ultrasonic waves and ozone for disinfection.

Specific gravity The ratio of the density of a substance to the density of a standard. The standard for solids and liquids is water with the density of 1 g/cm^3; for gases, the standard is air with the density of 1.29 g/L at standard temperature and pressure.

Spectator ion An ion in a reaction system that does not participate in the reaction.

Standard solution A solution with the precise concentration of the solute.

Stoichiometry Quantitative relationship between reactants and products in a chemical reaction.

STP The abbreviation for standard temperature (0°C) and pressure (760 mm Hg).

Structural formula A formula that shows the bonding in a molecule.

Supersaturated A solution that contains more dissolved solute than a saturated solution would contain under similar conditions.

Temporary hardness (carbonate hardness) Hardness caused by bicarbonates of calcium and magnesium.

Titration The technique where the concentration of a reagent is determined by reacting it with a reagent of known concentration.

Triprotic An acid capable of donating three protons per molecule.

Tyndall effect The dispersion of light by colloidal particles.

Valence electrons Electrons in the outermost shell.

van der Waals forces Forces of attraction between the nonpolar covalent molecules.

Vapor Gaseous state of substances that normally exist as solids or liquids.

Vapor pressure Pressure due to vapor of a confined liquid or solid.

Volatile Easily vaporized.

Zeta potential The magnitude of the charge at the boundary between the colloidal particle and the solution.

References

American Water Works Association. 1982. *Treatment Techniques for Controlling Trihalomethanes in Drinking Water.* Denver, Colo.: AWWA.

American Water Works Association Research Foundation. 1996. *Internal Corrosion of Water Distribution Systems.* Denver, Colo.: AWWA and AwwaRF.

Baylis, J.R. 1930. How to Avoid Loss by Pipe Corrosion by Water. *Water Works Engrg.,* 83.

Glage, W.H. and K.W. Kang. 1988. Advanced Oxidation Processes for Treating Groundwater Contaminated with TCE and PCE: Lab Studies. *Jour. AWWA,* 80:5:57.

Griffin, A.E. and R.J. Baker. 1959. The Break-point Process for Free Residual Chlorination. *Jour. NEWWA.*

Hurst, G.H. and W.R. Knocke. 1997. Evaluating Ferrous Iron for Chlorite Ion Removal. *Jour. AWWA,* 89:8:98.

Langelier, W.F. 1936. The Analytical Control of Anti Corrosion, Water Treatment. *Jour. AWWA,* 28.

McGuire, M.J. and M.K. Davis. 1989. Treating Water With Peroxone: A Revolution in the Making. *Water Engrg. and Mngmt.,* 135:5:42.

Metcalf, H.C., J.E. Williams, and J.F. Catska. 1966. *Modern Chemistry.* New York, London, Toronto, Sydney: Holt, Rinehart, and Winston.

Quick, F.J. 1965. *Introduction to Chemistry.* New York: McMillan Company.

Ryzner, J.W. 1944. A New Index for Determining Amount of Calcium Carbonate Formed by Water. *Jour. AWWA,* 36:4:474.

US Environmental Protection Agency. 1977. Water Supply Research, Office of Research and Development. *Ozone, Chlorine Dioxide and Chloramines as Alternatives to Chlorine for Disinfection of Drinking Water. State of the Art.* Cincinnati, Ohio: USEPA.

Index

Note: *f.* indicates figure; *t.* indicates table

V

W

Z